# Replenishment

# Replenishment

## How Deep Does It Go

Extended Version

## Donell Jackson

**Shift Elements LLC**

5830 E 2nd St, STE 7000 #9190

Casper, WY 82607

Send Feedback to Info@ShiftElements.com

Copyright © 2024 by Donell Jackson.

ISBN:    Softcover: 9798340479136

         eBook: 979-8-8690-6215-4



**Special Discounts for Bulk for Bookstore placement is available.**

**Please Contact Info@ShiftElements.com**

# Contents





Extended Version

Extended Version

Extended Version

Extended Version 

Extended Version

Extended Version

# Intro

A farmer slowly woke up, rubbing the sleep from his eyes. "Butch, you ready to start the day?" he said to his loyal dog, who was already waiting by the door. He walked down the stairs to the kitchen and prepared coffee to start his day.

Before heading out onto the porch with his dog, the farmer poured himself a cup of coffee. He then settled into his favorite rocking chair, ready to start his day. Butch lay down beside him, and they both gazed out at the field where the cattle grazed.

"Damn fog," the farmer muttered, squinting into the thick mist. "I can't even see the cows."

Butch let out a soft whine, as if sensing his owner's frustration.

Suddenly, the farmer's eyes narrowed. "Who's there?" he shouted, spotting a human type of figure

 Extended Version

walking through the fog where the cattle were, but there was no response. His grip on his coffee cup tightened. He signaled his dog. "Butch, go check it out."

Butch took off into the fog but couldn't find anyone. Lost and circling, he waited for further instructions from the farmer.

"Butch, come back here!" the farmer called, concern etched on his face.

When Butch returned, the farmer saw the figure again. "Go get 'em, boy!" he urged, sending Butch back into the fog.

This time, Butch ran straight toward the figure, but when he entered the fog, he ended up face-to-face with a cow. Butch's confusion deepened.

The farmer wasn't sure what was going on. He was wondering how Butch was unable to locate the figure that was walking through the fog. "Butch, come back!" he called, but Butch didn't return.

The farmer stood up, setting his coffee aside. "Butch, where are you?" he called, taking a step into his house to grab his shotgun that was placed right near the front door. Then he walked out of the house and straight toward the fog, looking around to see if he could see Butch. Butch was nowhere to be found, so the farmer tightened his grip on his gun and walked into the fog.

As the farmer walked further into the mist, his unease grew. He couldn't see anything, and the fog seemed to swallow him whole.

Suddenly, Butch shot past the farmer, running as if from something.

The farmer spun around, trying to follow, but Butch had already vanished into the fog. The farmer's fear grew tremendously, but he was lost in the fog, so he just stood still for a minute, looking

          Extended Version

around so he could try to find his way out of the mist. Frightened, he called out, "Butch, come back here," his voice echoing off the mist.

When Butch cleared the fog, he turned and started barking frantically.

The farmer's heart raced as he tried to make sense of what was happening, trying to walk toward the sound of Butch. And then, in an instant, everything changed. A burst of light illuminated the fog, and the farmer disappeared. His clothes were left behind, engulfed by the fog, as if he had never existed. A dark sphere hovered over his clothes and then flew off into the mist.

Butch continued barking, as if trying to alert someone to what had just happened, but there was no one to hear him. The fog swirled on, mysterious and unyielding.

Extended Version

# Replenishment

The fog slowly cleared as the sun rose higher in the sky. The cattle emerged, grazing peacefully, unaware of the strange events that had just transpired.

Butch, still barking, ran around in circles, as if searching for his missing owner. But the farmer was gone, vanished into the fog, leaving behind only his clothes. Butch had managed to pick up on the scent of his owner's clothes; standing there, sniffing them, he was confused about where his owner had gone. He continued to bark and run around the field, hoping to pick up on some clue but found absolutely nothing.



Extended Version

# Chapter 1

Deployed

Jimmy stood outside his father's office, taking a deep breath before entering. Jimmy had started reconnecting with his father. A lot has happened with his mother and father since the divorce. He had been promoted to the rank of commander. His father, Phillip, was now a detective-a career change that came after he had worked so hard, even though he lost his wife because of it. There was just so much going on in Jimmy's head, but he wanted to let his father know what was going on.

Jimmy walked in and saw his father working. "Hey, Dad," he said, trying to sound calm.

Phillip looked up from his desk, a mix of emotions on his face. He was curious why Jimmy had come to see him. "Jimmy, my son. I'm glad you came by." He got up from his desk, and they hugged in a tight, emotional embrace.

"I came but to let you know that I am being deployed. The war is finally here, and I'm going to miss you, Dad," Jimmy said, his voice cracking.

"Wow, that's a little surprise. I don't know if I'm happy about you being deployed, but I'll miss you too, Son. Be safe over there," Phillip replied, a little caught off guard.

Jimmy nodded, feeling a lump form in his throat. "I will, Dad. I promise. I don't know if I'm ready myself, but it's time to fight for our country," he said.

Phillip didn't know how to feel about it; he just wanted his son to return in one piece. He had already lost his wife, so if something happened to Jimmy, he wouldn't know what to do. He kept a straight face in front of Jimmy, but inside, it was

     Extended Version

eating him up. He didn't think that it would bother him that much. When he heard Jimmy say it, it had an impact he didn't expect.

Next, Jimmy headed to his mother's place. He hadn't been able to talk to her since he told her he was joining the army, and she hadn't spoken to him either. He wasn't sure what to expect. He didn't even know if she would even entertain a conversation. He was just going to expect the worst while hoping for the best.

When Jimmy arrived, Alice opened the door, looking hesitant. "Jimmy," she said, her voice barely above a whisper. She didn't know what to say, so she just decided to listen to whatever Jimmy had come to tell her. She felt a little guilty about not speaking to him all this time, but her pride wouldn't let her show her true feelings.

# Replenishment

"Mom," Jimmy replied, feeling a mix of emotions, "so I know you're not talking to me right now, but I have to let you know what's going on. So … umm …"

Alice stood silently, her eyes locked on Jimmy's, as she waited for him to gather his thoughts and reveal what was weighing on his mind.

"I don't know if you really care to know, but a war has started, and I'm being deployed," Jimmy said, looking nervous.

Alice still didn't say anything; she just stood there deep in thought. Her facial expressions couldn't hide her true feelings. She wasn't happy about the news at all. Even though she hadn't been talking to Jimmy, she still had a motherly love inside for him. She loved him more than he would ever know. He might have left her, but she thought about him every night. Thinking of him was the only thing

     Extended Version

that had kept her in the right state of mind during her divorce process.

They stood there for a moment awkwardly. Then Alice mustered up the courage to hug him. "Be safe," she whispered.

Jimmy hugged her back, feeling a sense of relief. "I will, Mom. I promise."

That was the most joy that Jimmy had felt in years from his mother. He felt an overflow of emotions. He wanted to open up and tell her everything that happened to him when they hadn't spoken to each other, but he didn't want to overdo it, so he kept his cool. He embraced her and hugged her for as long as she would let him. He could feel her heartbeat through the tightness of his grip. He felt like a baby in their mother's embrace.

Jimmy finally let his mother go after a minute or so, and his smile was big and undeniable. She didn't want to show her emotions, so she stepped back and closed the door behind her.

After visiting his parents, Jimmy returned home to spend his final moments with his wife, Jessica. From the instant they met, he knew she was the one. Now, as deployment loomed, he dreaded sharing the news, knowing it would shatter her heart. As soon as he made it back home, Jessica ran up to him and hugged him tight. He could hear her sniffing, trying to hold everything in.

Jimmy looked down at Jessica. "It's OK, honey," he said as he hugged her back.

Jessica cried as Jimmy held her close. "Come back to me in one piece. I need you … I can't lose you," she whispered.

"You won't. I will come back to you in one piece, my love," Jimmy replied, his voice shaking. "I promise."



Extended Version

# How Deep Does It Go

They stood at the door, holding each other tightly. Jimmy knew he had to leave but didn't want to let her go. Jimmy had a crazy life, and he finally had someone in his life who loved him unconditionally. He didn't ever want to let her down. What he had with Jessica was irreplaceable; she filled in all the empty voids his family could not fill.

"I love you with all my heart, and I will come back to you, no matter what it takes. I know you are scared, and so am I, but when I'm worried, I will keep you on my mind; when things seem to be impossible, I will think of you. You will be what I'm looking forward to coming back to," Jimmy whispered as he held Jessica.

"Aww … I love you too. You know you mean the world to me. I just want you to be safe, nothing more," Jessica replied.

# Replenishment

Jimmy took one last look at his wife, then turned and walked away. He got into the car, waving goodbye as they drove off. He knew he had to keep himself together for both of them, so his smile overpowered his true emotions.

Jessica stood in the doorway. As the car drove off, tears flooded her eyes. She started crying uncontrollably, her heart heavy with worry. She knew Jimmy had to go but couldn't help feeling scared.

As the car disappeared, Jessica took a deep breath and whispered, "Come back to me, Jimmy. Please come back to me."

Jimmy's deployment was a difficult time for everyone. Phillip, his father, was worried sick about him, but he tried to stay strong for his family's sake. Phillip had just been through so much that he wasn't ready for his son to be deployed. Alice, his mother, was a little quieter and more reserved, but her eyes

 Extended Version

betrayed her concern. Her relationship with her son wasn't where she wanted it to be with such a big event going on.

Jessica, Jimmy's wife, was a rock, but even she had moments of weakness. She wrote Jimmy letters every day, pouring out her heart to him, telling him how much she loved and missed him. Her thoughts drifted back to her early days with him, when love letters filled the distance between them. Revisiting those tender emotions brought solace, creating a sanctuary of memories that sustained her through the longest days and darkest nights.

Jimmy, meanwhile, was doing his best to stay safe and focused on his mission. He wrote back to Jessica whenever he could, telling her about his days and nights, the people he met, and the things he saw. He was getting ready to enter a zone that would be a lot more hands-on with his training.

# Replenishment

As the weeks turned into months, the letters became more sporadic. Jimmy was busy, and the internet connection was spotty at best. But Jessica didn't give up. She kept writing, hoping he would come home to her one day. And then, one day, the news came. Jimmy's unit was headed out to the front line, so she didn't know how to feel. She was in denial, and the long days had just gotten longer.

She didn't know how she would be able to deal with the new news. That was news that she would never be ready for-her husband being put in front of people who would be trying to kill him. She counted down the days, the hours, the minutes until Jimmy's return and if he would even return. It was something that would weigh down on her.

Phillip found out about the news, and all he could do was sit at his desk, his hands clasped over his head. He really didn't know how to feel. Not too long ago, he got the news that his only son was being

 Extended Version

deployed. Now, another devastating blow hit: his son was being sent to the front line. Losing a child was a father's biggest fear, and that would be something that his heart couldn't bear.

Alice took things a little harder than Phillip. Even though she had been cold toward Jimmy, when she heard about him being put in the middle of the war, she got so emotional that she cried, and this was the first time that she had cried since her divorce. She was really taking the news hard. She didn't eat for days. Her appetite was just not there, and not knowing what was going on with her son wasn't helping.

Jimmy reached his location and was locked and loaded. He knew that he was fighting for more than just his country. He would be fighting for his family and assuring them he would return home.

# Chapter 2

In Deep

Shots were screaming through the air like bullets from a hound dog. The team was holding the line while the Arabian army was bringing in the tanks. Planes were hazing the ground with shots. Dust was getting flipped up everywhere as missiles dropped from above like falling stars. Jimmy was radioing in for support.

"I need air support now!!"

The Arabian plane was coming back around to spray again, and one of the soldiers noticed the bullets as they dived down, creating a line of dust like Disney's way of doing a water show, and it was headed straight for his commanding officer; he screamed "Jimmy!" as he rushed over and pushed Jimmy out of the way. A second plane came in, following up and



Extended Version

shot a missile right in the direction that Jimmy was pushed.

Jimmy looked up and saw the missile coming in; he tried to jump out the way. It hit a couple of yards away and the force of impact pushed him so hard that his body flew uncontrollably right toward a demolished motorcycle. Before he knew it; his head smashed against its engine, and he blacked out.

Three months later, Jimmy was tossing and turning in his sleep as the nurse and his wife watched. His wife got up and grabbed his hand. He jumped because he could see the missile coming right for him. When he came to his eyes were hazy, he was unable to recognize where he was and then he noticed wires running everywhere. The streets of Iraq had shrunk down into the walls of his hospital room. His wife was overly excited that he had awakened.

# Replenishment

"Honey … Honey," Jessica said.

Jimmy looked at her and smiled, but he was in so much pain that he frowned a couple of seconds after.

"Are you ok honey?" Jessica asked, looking back at the nurse.

"My head hurts," Jimmy replied in a crackling voice, he didn't use his voice in 3 months.

"OK honey, the nurse will get you some pain medicine," Jessica said.

"My head is killing me."

"I know, honey. You shattered your skull while you were in the war. You had to get a metal plate put in your head because you had a severe brain injury."

"What!!!"

"Honey, calm down; it was put in to prevent any further brain damage."

"Where is everyone else?"

 Extended Version

# How Deep Does It Go

"The war is still going on. After your injury, you had to get shipped back immediately. You were in critical condition. The doctor said once you woke up, it was a good chance that you could fully recover."

"What do you mean 'woke up'? How long have I been asleep?" Jimmy asked, looking at his wife and the nurse simultaneously.

"Honey it's been 3 months."

"Are you serious? I've been in a coma for 3 months?" Jimmy replied as a tear swirled down his cheek. "Wow."

"Honey it's ok, I've been here every day. I couldn't stand you being in here like this," Jessica said, grabbing his hand.

Jimmy's voice was ruff as he tried to hold a conversation.

# Replenishment

"What about your job?" Jimmy asked in a mumbling high-pitched squeal.

"Honey it's ok, my job has been working with me through it all. What I need you to do is rest, and when your voice is working better, we can talk about everything. I'm just so happy to have your back," Jessica said, trying to hold in her tears as she kissed him on the forehead.

Jimmy closed his eyes and tried to get some rest. Jessica looked up and silently thanked God as she sat down in the chair beside the bed, holding Jimmy's hand, not wanting to let go. She had been through so much during this time. This was one of the hardest things that she had to do.

Watching him day in and day out just lying in the hospital bed almost drove her crazy. She hoped he would awake soon, so she could see his beautiful brown eyes again. She was close to running out of vacation time at her job. She had saved it up to take

 Extended Version

time off when Jimmy was back from the war, so they could spend some time together and catch up for all the time that he was away. Jimmy being in the hospital wasn't what she had in mind, but it needed to be done-and what a better time than when he needed her the most.

She spent sleepless nights weeping, as the moon illuminated a bright white light that glared through her tears, which never left her alone, not even through the beeping machines that imitated the beats of his heart. The nurse kept a fresh box of tissues beside the bed which had become her knight in shining armor that guarded her against the darkest dragon of them all-a lost husband that wasn't lost in body but lost in a dream which seemed like an eternal slumber.

A few years ago, Jessica had given up smoking, but dealing with this situation brought her skeletons back out of the closet. Stress took over her mind and suffocated her from the inside out, pulling and pulling at her until she gave in like a virgin on prom night. Everyone has a breaking point, and these last 3 months surely didn't take any prisoners when it came to her situation.

She noticed–when things got bad–just how many people seemed to fade out of her life. Those whom she thought were her friends were nowhere to be found after the first month had passed while Jimmy was in a coma, she needed someone to lend her a little money so she could cover some bills. Jimmy had money, but she didn't have his information and her money wasn't covering everything that needed to be taken care of.

Through everything, she sorted out how many true friends she had. Jimmy's parents helped out a lot. She took the money, because she had no one else to turn to, her parents were the last people she

 Extended Version

wanted to help. Nothing was ever just–because it was always a loan that had to be paid back, or they would treat her as if she didn't exist. So, for things not to become complicated, she dealt with everything on her own, and that's how she has always been. Thinking about all these things as she looked at Jimmy made them worth it. She leaned over, kissing Jimmy on the forehead as he lay there, she whispered, "I love you and I'm so happy you're back with me. I really missed you."

She was so excited that she just leaned back in the chair and watched Jimmy sleep for hours. She didn't feel alone anymore, nor did she feel as if she were taking on the world all by herself. She felt as if the world had lifted off her shoulders.

Extended Version

# Chapter 2

Love at First Sight

Flashback:

Jimmy was being deployed to Germany, a place he had always wanted to visit. Having always prioritized his mother's care, he never thought he'd be able to do this anytime soon. But that was about to change–and under less-than-ideal circumstances. This was something that Jimmy would still be able to check off his bucket list.

Jimmy received the information about his deployment a few days ago, detailing when and where he would be heading. He lived on base, so he didn't have to check in with anyone.

Jimmy had finally made it to Germany. He stepped off the plane, his heart racing with excitement and nerves.

            Extended Version

"Welcome to Germany, soldier!" a gruff voice shouted.

Jimmy saluted, trying to look confident. "Thanks, Sergeant," he replied. He looked around, taking in the unfamiliar sights and sounds. "Wow, I can't believe I'm finally here," he said.

"I know, right?" a soldier said, slapping Jimmy on the back. "You're gonna love it here. I'm Sarge, by the way."

Jimmy smiled, trying to fit in. "Thanks, Sarge. I'm Jimmy."

"OK, Jimmy, this should be fun," Sarge replied.

"Yeah," Jimmy said, not too sure what to make of Sarge's comment.

The sergeant looked around and shouted, "Solider James, can you show Jimmy to his quarters?"

# Replenishment

"Yes, sir," James replied. "This way, Soldier," he said, looking at Jimmy.

Jimmy followed James to his quarters.

"So, how do you like it here?" Jimmy asked, curious about James's response.

"It's OK, just like any other place. You come in, do your job, then move on to the next," James replied.

"I see," Jimmy said.

"You just have to learn not to get too comfortable in one place because you may not be there long enough to take a shit," James said.

"Well, I guess I better hit up the bathroom right away," Jimmy said with a laugh.

"I see you have a sense of humor," James said with a laugh, even though he was trying to keep a straight face.



Extended Version

# How Deep Does It Go

They finally reached Jimmy's quarters, and as soon as Jimmy walked in, Sarge looked at him with a grin.

Jimmy turned and whispered to James, "So, what's his problem?"

"A little word of advice: steer clear. You don't want to be mixed up with that guy," James replied.

Jimmy nodded his head as he turned back and looked at Sarge. "Noted," he said.

As the days passed, Jimmy settled into his new routine, but Sarge and a few others kept giving him a hard time.

"Hey, Jimmy! You're still alive?" Sarge teased.

Jimmy rolled his eyes. "Yeah, barely."

One soldier, Mike, took Jimmy under his wing. "Hey, so what's your name?" he asked.

"Jimmy. Yours?" Jimmy replied.

"You can call me Mike," Mike said.

Extended Version

# Replenishment

"OK, cool," Jimmy said; he noticed the other guys laughing at him.

Mike turned to Jimmy. "Yeah, it's tough, but don't worry, Jimmy. We've all been there. Just ignore them."

"I try to, Mike, but it's hard," Jimmy said.

Mike nodded sympathetically. "I know, but you'll get used to it. Just focus on your training, and you'll be fine."

"Hey, Jimmy! You're still alive?" Sarge teased, chuckling.

"Wouldn't you like to know?" Jimmy replied, rolling his eyes.

"Come on, Jimmy! You can do better than that!" another soldier chimed in.

"Yeah, you're not exactly the sharpest tool in the shed, are you?" Sarge sneered.

Jimmy sighed, trying to ignore them.

"Hey! Knock it off, guys!" Mike stepped in.

 Extended Version

"Mind your own business, Mike," Sarge growled.

"Make me," Mike said firmly. "What you're doing isn't called for. We should be supporting each other, not tearing each other down."

The other soldiers looked sheepishly, but Sarge continued to taunt Jimmy. "Oh, I'm shaking in my boots, Mike. You think you're so tough defending the newbie?"

Mike stood his ground. "Call it what you want, but we are all on the same team."

Jimmy nodded, grateful for Mike's support. "Thanks, Mike. I appreciate it."

Sarge snorted. "You're just making things worse, Mike. Jimmy needs to toughen up if he wants to make it."

# Replenishment

Mike shook his head. "That's not how we do things, Sarge. You were just like Jimmy when you first started. So, cut the shit."

The tension was palpable, but Mike's words seemed to have an impact on the other soldiers. They nodded in agreement, and Sarge backed off, muttering under his breath.

Jimmy felt more confident as the training continued, knowing that Mike had his back. And when they took a break, Mike sat down next to him.

"Hey, Jimmy, I don't know what Sarge's problem is. He's just trying to get a rise out of you," Mike said.

Jimmy smiled. "Yeah, I know, bro. I appreciate it. I won't pay him no mind."

Mike patted Jimmy on the back. "That's my boy, Jimmy. We're all in this together. I got your back."

 Extended Version

# How Deep Does It Go

A few days later, the guys decided to head into town. "Come on, Jimmy! Let's show you the real Germany!" Mike said.

Jimmy grinned, excited to explore.

As they walked through the streets, Jimmy marveled at the architecture and the sounds of the language. "Wow, this is crazy," he said.

"I know, right?" Mike said. "Germany is a great country. You'll love it here."

That's what Jimmy had been waiting to hear. He was more excited now than ever and was ready to see everything.

"Today's the day we explore Germany!" Mike said, grinning.

Jimmy's eyes lit up. "I can't wait! I've heard so much about this country."

Mike chuckled. "You're in for a treat, my friend. Germany has a lot to offer."

Extended Version

# Replenishment

As they walked through the streets, Mike pointed out historic landmarks and shared stories about the culture. Jimmy listened intently, soaking up every word.

"This is pretty cool!" Jimmy exclaimed, gazing up at a stunning cathedral. "The architecture is incredible."

Mike nodded. "Yeah, Germany has a rich history. And the food? Forget about it. You're in for a treat."

They stopped at a traditional German restaurant, where Mike ordered a variety of dishes. Jimmy's eyes widened as the food arrived.

"What's this?" Jimmy asked, pointing to a sausage.

"That's a Bratwurst," Mike said. "Try it; you'll love it."

Jimmy took a bite, and his eyes lit up. "OK, it's good. I see you have a taste for good food!"

                    Extended Version

Mike laughed. "Told you. And try the sauerkraut; it's a classic."

As they ate, Mike told Jimmy about the different regions of Germany, the festivals, and the people. Jimmy listened, enchanted by the stories.

After lunch, they strolled along the Rhine River, taking in the breathtaking views. Jimmy couldn't believe how beautiful Germany was.

"Thanks for showing me around, Mike," Jimmy said, genuinely grateful. "I really like it here. It's really different from the states."

Mike smiled. "Yes, it is. I knew you would like it. And I appreciate you for letting me show you around."

As the sun began to set, Mike and Jimmy grabbed some traditional German beer and sat down at an outdoor cafe.

# Replenishment

"Prost!" Mike said, clinking his glass against Jimmy's.

"Prost!" Jimmy replied, taking a sip of the cold beer.

As they sat there, watching the stars come out, Jimmy turned to Mike with a smile. "Thanks for today, bro. I really needed this. Training has been tough, and dealing with Sarge hasn't made it any easier."

Mike nodded. "Anytime, Jimmy, and I understand. Sarge is a jerk, but I've got your back, bro. That's what friends are for."

As Jimmy continued to take in the surroundings, his gaze fell upon Jessica, a beautiful American tourist on vacation with her friends. She looked up at Jimmy, and they locked eyes. Jimmy smiled so big that he couldn't even hear what Mike said. Before he realized it, Jessica had walked right up to him.

 Extended Version

"Hi, I'm Jessica," Jessica said, smiling.

Jimmy's heart skipped a beat. "Hello, I'm Jimmy. Nice to meet you."

Mike had got the hint. "Well, Jimmy, just meet us at the bar later. Most of the guys will be there. I'll shoot you the address," he said.

"OK, bro. I appreciate you. I'll see you later," Jimmy replied. Looking back at Jessica, he said, "Sorry, this is my first time in Germany. So, one of my soldier buddies was showing me around."

"Oh, this is your first time in Germany. It's mine also," Jessica said.

"Word … That's what's up," Jimmy said.

They continued to talk, and before they realized it, they had been talking for hours, laughing and joking. Jimmy felt like he'd found a kindred spirit.

Extended Version

# Replenishment

Realizing it was getting late and he had to meet up with his comrades, Jimmy said, "Jessica, it was a pleasure meeting you. I really enjoyed our time. I was hoping that we could continue this another day before you leave." He was nervous, but he still decided to ask, "So, is it OK if we exchange numbers?"

Jessica paused for a moment, as if she had to think about it. "Sure ... I would love that."

Jimmy took Jessica's number and hurried off to meet the guys at the bar.

On getting to the bar, Jimmy saw Mike and a few other guys. Sarge was there also; Jimmy wasn't too happy to see him there. Even though Sarge was there, Jimmy tried to make the best of it. He was just hoping that Sarge wouldn't be a problem.

As the night wore on, Sarge started causing trouble at the bar. "You locals are so stupid," he sneered.

 Extended Version

"Hey, Sarge, calm down!" Mike tried to intervene, but it was too late.

A scuffle broke out, and Jimmy found himself in the middle. "Stop it, Sarge!" Jimmy shouted. "You're going to get us all in trouble!" He was hit and ended up knocking out a few guys.

Mike grabbed Jimmy to let him know that it was time to go. "Jimmy, I see you have a little scrap in you."

"Yeah, a little," Jimmy said with a smile.

"Let's get out of here, man. I hope we don't get in trouble for Sarge's mess. Good to see that you have your brothers' backs, even when they are wrong," Mike said.

"I know. I tried to de-escalate the situation, but, man, the nerves of that guy!" Jimmy replied.

Extended Version

# Replenishment

Jimmy and Mike scrambled out of the bar and headed back to their barracks, praying the whole way back.

The next day, the head commander called Jimmy and the others into his office. "What happened last night?" he demanded. "Sarge, did you have anything to do with this?"

Sarge just sat there, quiet.

"So, you can't hear me? I swear, if you started that last night, you will be on the first plane out of here," the commander said.

"But Commander …" Sarge replied.

"You heard what I said, Soldier!" shouted the commander, looking around. "What about you, newbie?" asked the commander, looking at Jimmy.

Jimmy took a deep breath. "It was my fault, sir. I tried to de-escalate the situation."



Extended Version

Sarge glared at Jimmy. "No way it was his fault! Sorry, Commander," he said, but Jimmy stood his ground.

The commander looked at Jimmy with newfound respect. "Well, Jimmy, it takes a big man to take the blame. You're a true leader."

As they left the commander's office, Mike patted Jimmy on the back. "You did the right thing, Jimmy. We've got your back."

Sarge looked sheepish. "Yeah, sorry about that, Jimmy. My bad."

Jimmy nodded, feeling a sense of camaraderie. "No problem, Sarge. We're all good."

A few days later, the smoke had cleared up a little back on base, and Jimmy was ready to see Jessica again before she left; there would be no telling when he would see her again. Jimmy texted Jessica to see if she was free to meet up. She had

Extended Version

some time, so they planned to meet later that night–and eventually did.

Jimmy and Jessica both smiled from ear to ear as they approached one another; their connection was still strong.

"I can't believe how much we have in common," Jessica said.

Jimmy grinned. "I know; it's crazy. I feel like I've known you my whole life."

Jimmy felt like he was on cloud nine as they walked through the streets. He had never felt this way about anyone before. "I'm so glad I met you, Jessica," he said.

Jessica smiled, her face flushing. "I'm glad I met you too, Jimmy."

Their connection only intensified as they explored the town together, laughing and joking. Jimmy felt like he had found his soulmate. He was

 Extended Version

just hoping that Jessica felt the same way. "I can't believe how lucky I am to have met you," he said.

Jessica smiled, her eyes shining. "I feel the same way."

Jimmy was ecstatic to hear that coming from her.

They sat on a bench, watching the sunset. Jimmy had grown a large liking to Jessica. "I never thought that I would come to Germany and find a woman like you," he said.

Jessica's face lit up. "You are really sweet. You seem like a good man yourself."

Jimmy really liked being in Germany, and he had now found his heart with Jessica. He was grateful for the chance to serve his country and to find a wonderful woman in the process. He had been stuck in the house taking care of his mother so much that he didn't even think his life would shape up the

Extended Version

way it did. He despised his mother for the longest time. He missed out on so many things and experiences because of her being sick.

"I'm so glad I got deployed here," Jimmy said, smiling at Jessica. "I never would have met you otherwise."

Jessica smiled back; her eyes sparkling. "I'm glad too, Jimmy. I couldn't imagine being with anyone else."

# Chapter 3

Real Love

Jimmy had been in the military for a few months now. He was getting better with his training. He was also fitting in with the fellas. He and Mike were like best friends at this point. From the moment they met, Jimmy and Mike discovered they shared a surprising number of common interests. Mike had been looking out for him since he arrived in Germany, and Jimmy was grateful for his support. Jimmy never had a brother or any other family member that he was close to, so he chose to build that bond with Mike.

Even though he had a strong bond with Mike, Jimmy found himself growing tremendously close to Jessica to the point that he would sit at his desk, pen in hand, staring at the blank piece of paper for hours, not knowing where or how to start or even what to say. Jimmy had been so cut off from the rest of the world that he found himself uncertain about how to step through the door that had finally swung open. It was as if Jimmy had become stagnant and unable to fully open up to new people.

Even though Jimmy was stationed in Germany, training hard every day, his mind was always on Jessica. Ever since that day they met while she was on vacation in Germany, their connection had grown over the last few months.

Jimmy had no idea he would come to Germany and run into the most beautiful, caring, driven, and dedicated woman he had ever met. He would always think, "What if I had never left home? I would have never had a chance to meet Jessica." She has shown him how much more there is to discover and appreciate in this world. He was just so happy that

 Extended Version

he didn't miss his chance to experience all the new adventures that he had been missing out on most of his life.

Jimmy finally stopped reminiscing and started to write Jessica a letter:

*I can't believe we ran into each other like that. It was like fate brought us together. I really enjoyed spending time with you. I just would like to thank you for blessing me with your time. I know there were probably a million other things that you could have done, but you chose to hang with me. I really appreciate that. When I first made it to Germany, the guys gave me such a hard time, being the new guy and all, so I was happy that I was able to get away from it all and enjoy Germany with a beautiful woman.*

Extended Version

# Replenishment

Jimmy paused for a moment, wondering if he should get a little deeper with her and tell her about his family. He put the pen back to the paper and continued:

*So, I know that I haven't talked much about my family. I feel comfortable with you, so I would like to tell you a little. My father's name is Philip, and he is a detective. He and I have an OK relationship, even though he wasn't around much because he had to work a lot. Our relationship could have been better, but it is what it is. At least we talk, so I'm OK with that. I never really experienced what it's really like to have a father who was around for all of life's unexpected experiences. Even though he hasn't really been there, he is always fighting to show me his worth as a father.*

 Extended Version

# How Deep Does It Go

Jimmy paused once more before continuing because his mother's story was a lot different. He cracked his back and continued once more:

*My mother, on the other hand, isn't talking to me right now. Most of my teenage years was about taking care of my mom. When she got a little older, she started to become sick a lot. So, I became the caregiver. My father wasn't around, so I pretty much became the father and the son. I found myself catering to my mother's needs all the time, leaving me with little time to do anything else. I was never able to go on field trips, parties, or anything else that teens did at that time. I felt like I missed out on so much in life during that time. But after a while, I started to grow tired of being my*

# Replenishment

*mother's caregiver. I met a recruiter who opened my eyes, and I ended up having to make a tough decision to join the army. So, I'm here, even though it ruined my mom and I's relationship. She hasn't talked to me since I told her that I was going to join the army.*

Jimmy placed the pen on the desk so that he could pull himself together because he really got emotional talking about his parents-and, more so, his mother. He had so much anger built up inside, so he tried to avoid anything that could trigger him. Ever since he made it to Germany, he had to overcome situation after situation. The only thing that had been helping him through it all was Jessica, and the letters that they had been writing back and forth had been like therapy for him.

 Extended Version

# How Deep Does It Go

Jimmy had written enough this time around, so he ended the letter, sealed it up, and wrote the shipping address so that he could send it off to Jessica, while his anticipation kept him anxious for his return letter.

A week or so later, Jessica received her letter from Jimmy. When she pulled the letter out of the mailbox and saw that it was sent by Jimmy, her whole face lit up, and she felt all bubbly on the inside. Their bond grew stronger and stronger each day. The letters were their best option when it came to communication. It was a way to learn a lot more about each other. They could just sit back and take in everything they chose to share that time.

Jessica wrote back:

*I know exactly what you mean. I was feeling so lost and alone on my vacation, and then I saw you and Mike walking*

Extended Version

*down the street. It was like a breath of fresh air. Even though my friends were there, I was still a little lonely. They both have husbands at home, and I don't even have a pet. I used to always dream about meeting the right person, but the dating game has been brutal. Most men don't have their stuff together but want you to. Some lie about being single while having a whole wife at home. I tried dating apps, but those are the worst. I only ended up trying it because I was referred by a friend.*

After Jessica had finished talking about herself, she decided to put on a more serious face to respond to Jimmy's family situation. She didn't really know how to relate, seeing that she had both parents in her life. She was going to just be

 Extended Version

straightforward about the situation and hoped that she didn't overstep boundaries because the last thing she wanted to do was make Jimmy feel uncomfortable. She started to write again:

*Wow, I'm sorry to hear that you aren't as close to your parents as you would like to be. I'm sorry that your mother isn't talking to you right now. I hope she comes around because life is too short. We must love those who are in our lives while we can. We all know that life isn't promised tomorrow, so I hope she can get over your decision. Sounds like you needed a change in your life. I'm glad that you and your dad still have somewhat of a relationship.*

After all the awkward conversations, the letters they were writing had become a thing. Jimmy smiled as he read Jessica's letter. He had been thinking about her nonstop since the day they met. As he wrote, he told her about his training, about how he was hitting his shot targets dead center and getting praised by his comrades Mike and Sarge. He told her about the bittersweet relationship he was building with Sarge, who was tough on him but also had a soft spot.

Jimmy also told Jessica that it was okay with his parents; he explained how tough it was but assured her he was moving forward and shared how these letters had been very therapeutic for him throughout this process. But most of all, he wrote about how much he missed her and wanted to see her again, hold her in his arms one day, and kiss her.

He was really into Jessica; he shared some of his most personal stuff with her. Stuff that he hadn't really talked about with anyone. Jessica had become a safe place for Jimmy; he loved it, but it scared him

       Extended Version

also. He had never had anyone outside his mother and father whom he truly cared about.

Jessica wrote back, telling Jimmy about her family, her favorite color (blue), her favorite foods (pizza and ice cream), and the things she enjoyed doing most in life (hiking and reading). That's why she enjoyed going back and forth with him with the letters. Jimmy gave her a chance to read, and it was something that she was really interested in. She had really learned so much about Jimmy in a short amount of time. Family was a big thing to Jessica, and she knew that Jimmy was having a hard time with his, so she tried to keep it brief. The last thing she wanted to do was trigger him.

Jimmy devoured every word, feeling like he was getting to know Jessica better and better with each letter. Whenever she spoke about her family, he took it in but didn't really say too much on the

Extended Version

topic. He just accepted that her relationship with her parents was better than that of him and his parents.

He wrote back, telling her about his day, new endeavors with the guys in his unit, and how much he missed her and couldn't wait to see her again. Jimmy had started developing deep feelings for her, unsure how to stop his emotions. This was a new experience for him, and he was just taking all this one day at a time.

As the weeks turned into months, Jimmy found himself looking forward to Jessica's letters more and more. Her letters had started becoming an addition to him; he craved them like a crackhead waiting on his next hit. He felt like he was falling in love with her, even though they weren't seeing each other face to face. But he didn't care. He knew he wanted to spend the rest of his life with her and was willing to do whatever it took to make that happen.



Extended Version

## How Deep Does It Go

Jimmy was stationed in Germany, but he just stayed consistent with keeping in contact with Jessica until he could figure out how they could be together for good. He really could see her being his wife one day. He was just trying to figure out how to ask her to be his forever. Since they could talk to each other through letters and pretty much tell each other anything, he decided to just throw it out there in his next letter:

*Dear Jessica,*

*I know we only spent a little time in person, but I feel like I know you better than anyone else in the world. Throughout this process of getting to know each other, I learned so much about you, and I have also shared some stuff that was dear to my heart. Things that I have not shared with anyone. Over the*

Extended Version 

*course of getting to know you, I have discovered that we have more in common than I could have ever expected. You have not only become an important person in my life; you have become that special person in my life. You're my best friend; you're who I look to for advice. You have been my safe place, my happiness, and most of all, my peace. I never, ever thought about having my own family, but since we have been talking, I've thought about that and so much more. I thought about us building together, loving each other, and attacking this world as a power couple. You have made me look at life totally differently and opened my eyes to things that I have never even thought about, let alone manifesting those things into reality. I know it's only been seven*



Extended Version

*months, but when you know, you know.*

*So, I hope I'm not being too forward, but*

*I want to spend the rest of my life with*

*you. Will you marry me?*

Jimmy sealed the letter in an envelope as his heart raced with excitement. He couldn't wait to see her response. He could only hope that it was the response that he wanted. He had never even asked a woman to be his girlfriend, let alone marry him. So, this would certainly be a new chapter in his life if it came true. There were so many things running through his head: *What next? Do I want kids? How would I tell my parents? Would they accept her?* He had no clue how this would all pan out.

Days passed, and nights seemed longer. As the end of the week approached, Jimmy's anticipation grew. He even shared a few conversations with Mike about what he had done. All he could do was try to talk it out while waiting. He knew the letter wouldn't get there overnight, but he hoped it would somehow. He didn't want to get to the point that the waiting was starting to get to him, so he pushed harder during training and joined in on some games with some of the guys. After a while, the guys even realized how impatient he was getting.

About three weeks later, Jimmy was out in the field, training hard. It seemed like so much time had gone by since he sent the letter off that his anxiety was down to a three instead of a ten it had been since he first sent the letter off. As soon as training was over that day, he headed straight to the mail-receiving area. And there it was, a letter from Jessica.

Jimmy hurried back to his quarters so he could sit down and read the letter he had finally received. As soon as he flopped down on his bed, he ripped it

 Extended Version

open, his heart racing with excitement. Mike was on his bunk and could see that the day had come.

Jessica wrote:

*I really wasn't expecting to receive a letter like this. First, I'll start off by saying that you have really been a blessing in my life. You have taught me so much, and I've grown to love you over time. I've never met a man quite like you. You have made me so comfortable in so many ways, and I've opened up to you like I've never opened up to any other man. You have shown me that there is someone out there for everyone, and you are my person, and I love you, and I'm so happy that I met you when I came to Germany.*

*I really wasn't in a place where I was looking for anyone, but when faith*

Extended Version

*happens, sometimes you just go with it. It has been a journey getting to know everything about you. I can say that I never thought that I would do a long-distance relationship. In this case, it's more than just a long-distance relationship.*

Jimmy laughed and continued to read:

*I've had so many bad relationships that I almost didn't want to take a chance on a good one. I had so many thoughts, and I didn't know if I really wanted to move forward with us. I just want to say that I'm happy that I took a chance with you. You are a wonderful man. You're hard-working, persistent, loving, funny, and you have a big heart. So, I would be a fool if I said no. I would love to spend*



Extended Version

*the rest of our lives together. Yes, I will marry you.*

Jimmy's face lit up with excitement.

Mike looked over. "So, what did she say, man?" he asked, even though he knew by the look on Jimmy's face.

"She said yes!" Jimmy exclaimed.

"Congratulations!" Mike said.

"Thanks," Jimmy replied.

"So, I guess I will be a godfather one day, huh?" Mike asked.

"So, you're just going to jump the gun, huh?" Jimmy said.

"Ahh, what's the point in beating around the bush?" Mike replied.

All Jimmy could do was laugh.

"So, what's your next step," Mike asked.

Extended Version

"Not sure yet," Jimmy said, with a clueless look on his face.

"I'm sure you will figure it out," Mike said.

"Yeah, I hope so. She's my first everything, so I may need a little help," Jimmy said.

"What do you mean your first everything?" Mike asked.

Sarge walked in. "What did I miss?" he asked. "Did you guys hear about all the missing people back in the States?"

Jimmy took a deep breath. He didn't really want to share the information with Sarge.

Mike looked shocked. "Missing people?" he asked.

"I thought we were cool, Jimmy?" Sarge said.

"We are; just was talking with Mike about a few things," Jimmy replied.

"OK, so what's going on?" Sarge asked.

Mike looked at Jimmy. "It's cool; I got you, Jimmy," he said, walking out.

"OK, I appreciate it," Jimmy replied, lying back on his bed.

Sarge had a look of confusion on his face. "OK, I get it," he said.

Jimmy felt like he was on top of the world as he lay on his bed. He knew he had found his soulmate and was willing to do whatever it took to make her happy. He grabbed his notepad to write back; he knew their love would last a lifetime. And he couldn't wait to see what the future had in store for them.

# Chapter 4

Wedding Bells

Five months later, Jimmy sat at his desk, pen in hand, writing a letter to Jessica. They were thousands of miles apart, but their love had only grown stronger, despite the distance.

*Dear Jessica,*

*I was thinking we should start planning our wedding. I know we're far apart, but I can't wait to spend the rest of my life with you. You have brought so much happiness into my life, and I can't wait for you to be my wife. I was so happy the day that you told me that you would be my wife, and I've been thinking about you nonstop.*



Extended Version

## How Deep Does It Go

Jessica received the letter a few weeks later, and her heart skipped a beat as she read Jimmy's words. She wrote back:

*I'd love to start planning our wedding! But where should we have it? Your hometown or mine?*

Jimmy received her letter and wrote back:

*I was thinking, since my mom isn't speaking to me, it would be better to have it in your hometown. Plus, it would be closer to your family, and they could attend more easily. Most of my friends are here. The only family that would actually attend is my dad. I have a few cousins that I will ask, but you have more family members that would probably attend compared to me.*

Extended Version

# Replenishment

Jessica wrote back:

*That makes sense. My family would love for it to be here. Are you planning on having Mike be your best man? How many groomsmen are you planning on having?*

Jimmy smiled as he read Jessica's letter. He knew Mike would be honored to stand by his side on his special day. He wasn't sure about the groomsmen, but he would surely ask around. He has made a few friends since he has been in the service, so he was pretty sure he could find a few who liked him enough to be a part of his wedding.

The next day, Jimmy was ready to talk to Mike about the whole thing and see what he thought about everything. He saw Mike walking and stopped him to have a quick conversation.

 Extended Version

"Hey, Mike, guess what?" Jimmy said as Mike walked into the room. "Jessica and I are planning our wedding, and I was wondering if you would be my best man?"

Mike grinned. "That's awesome, Jimmy! It would be my honor to be your best man."

Jimmy nodded, feeling a sense of gratitude toward Mike. "Thanks, man. I really appreciate you, and I honestly don't know what I'd do without you. You have really helped me out when I was in some tough situations. You are a true friend."

That really put a smile on Mike's face. "No problem, man. It was nothing."

They finished chopping it up, and Jimmy went back to plan his wedding with his fiancée. Jimmy and Jessica decided to get married in three months, when Jimmy would return home during his break. He would fly out to Maryland, where they had both

# Replenishment

decided the wedding would be, and they would tie the knot in front of their friends and family.

In his last letter before the wedding, Jimmy wrote:

*I promise to love and cherish you forever and always. I talked to Mike also, and he has been helping me figure out the right steps to take in this process, seeing that I'm new to stuff like this. You're my first girlfriend, so planning a wedding is a little out of my element, but things are coming together quite well.*

Jessica smiled as she read his words, her heart filled with joy. She wrote back:

*I promise to love and support you through all of life's ups and downs. I know this is new for you. Shoot, it's new for me also. We are both just taking this*



Extended Version

*one day at a time. I'm happy that you have Mike there for you. I have my friends here for me, so I'm happy that you have someone there for you also.*

Three months later, the day finally arrived, and Jimmy flew to Maryland to meet Jessica. It had been so long since they had seen each other, so when they did, they smiled so hard that their faces started hurting. They started to slowly walk up to each other and hugged each other so tightly they didn't let go for a while.

*** ***

Jessica and Jimmy's wedding day had finally arrived, and the atmosphere was electric. The venue was transformed into a breathtakingly beautiful setting, with white and blue decor that seemed to dance in

Extended Version 

the soft light. The air was filled with the sweet scent of flowers and the gentle hum of excitement.

As the guests began to arrive, they were greeted by the elegant sight of white linens, blue glassware, and delicate white flowers that adorned the tables. The ceremony area was equally stunning, with a beautiful white aisle runner leading up to a gorgeous archway draped in flowing white fabric and adorned with blue delphiniums.

Despite the joyous occasion, a subtle yet unmistakable shadow fell across the gathering: Jimmy's mother was absent. She had refused to come, a decision that had weighed heavily on Jimmy's heart. However, he was determined not to let it dampen his spirits, and he was surrounded by loved ones who were eager to celebrate with him.

Jimmy's father, Philip, beamed with pride as he took his seat at the front of the ceremony area. He was joined by Jimmy's best man, Mike, who had

          Extended Version

been by Jimmy's side since they met when Jimmy joined the army, and their bond had remained unbreakable. Speaking of which, Jimmy's military buddies had turned out in force, their uniforms a testament to their camaraderie and shared experiences. They mingled with the other guests, swapping stories and laughter as they awaited the ceremony.

As the guests took their seats, the room fell silent in anticipation. Jessica's family was seated on the left, resplendent in their finery, while Jimmy's guests filled the right side of the venue. The air was thick with emotion, and the sense of community was palpable.

At the back of the room, Jessica's bridesmaids were busy making final adjustments to her stunning white gown. Her hair was styled to perfection, and her makeup was subtle yet radiant. She looked like a true princess, and her smile lit up the room.

Meanwhile, Jimmy was calm and collected, his military training evident in his poised demeanor. He adjusted his cufflinks, took a deep breath, and prepared to meet his bride at the altar.

As the music began, the room erupted into a flurry of activity. The bridesmaids made their way down the aisle, followed by Jessica; her heart raced with excitement as she prepared to walk down the aisle. She took a deep breath, her eyes fixed on Jimmy, who awaited her at the altar.

"Ready, Jess?" her father asked, offering his arm.

Jessica nodded; her smile was radiant. "I'm ready, Dad."

As Jessica reached the altar, Jimmy's eyes locked onto hers, his face filled with love and adoration.



Extended Version

"Dearly beloved," the pastor began, "we are gathered here today to witness the union of Jimmy and Jessica in marriage. God, we come before You today to witness the union of Jessica and Jimmy. Lord, bless this couple and their marriage with your love and guidance. Jessica and Jimmy, you have come here today to make a public commitment to one another. As you stand before us, remember that marriage is a sacred bond between two people. If you have prepared vows, you may now recite them," the pastor said.

Jimmy and Jessica exchanged their vows, their voices trembling with emotion.

"I promise to love and cherish you through thick and thin," Jimmy said, his eyes never leaving Jessica's face.

Extended Version

# Replenishment

"I promise to love and support you through all of life's ups and downs," Jessica replied, her voice barely above a whisper.

"Jessica, do you take Jimmy to be your husband, to love and cherish him through all the joys and challenges of life?" the pastor asked.

"I do," Jessica replied.

"Jimmy, do you take Jessica to be your wife, to love and cherish her through all the joys and challenges of life?" the pastor asked.

"I do," Jimmy replied.

"Jimmy, please place the ring on Jessica's finger and repeat after me. I, Jimmy, take you, Jessica, to be my wife, to have and to hold from this day forward, for better or for worse, for richer or for poorer, in sickness and in health, to love and to cherish, till death us do part, according to God's holy ordinance; and thereto I pledge you my faithfulness," the pastor said.

 Extended Version

# How Deep Does It Go

"I, Jimmy, take you, Jessica, to be my wife, to have and to hold from this day forward, for better or for worse, for richer or for poorer, in sickness and in health, to love and to cherish, till death us do part, according to God's holy ordinance; and thereto I pledge you my faithfulness," Jimmy said.

"Jessica, please place the ring on Jimmy's finger and repeat after me. I, Jessica, take you, Jimmy, to be my husband, to have and to hold from this day forward, for better or for worse, for richer or for poorer, in sickness and in health, to love and to cherish, till death us do part, according to God's holy ordinance; and thereto I pledge you, my faithfulness," the pastor said.

"I, Jessica, take you, Jimmy, to be my husband, to have and to hold from this day forward, for better or for worse, for richer or for poorer, in sickness and in health, to love and to cherish, till death us do part, according to God's holy ordinance; and thereto I pledge you my faithfulness," Jessica said.

"By the authority vested in me, I now pronounce you husband and wife. Jimmy, you may kiss your bride," the pastor said.

Jimmy and Jessica shared their first kiss as husband and wife, the crowd erupting in cheers and applause.

"I love you," Jimmy whispered, his lips still pressed against Jessica's.

"I love you too," Jessica replied, her eyes shining with happiness.

Together, they walked down the aisle, Jessica's family and friends rising to their feet as they passed. Jimmy's father, Philip, beamed with pride, surrounded by his son's army buddies.

After the ceremony, the guests made their way to the reception area, where a sumptuous feast awaited. The reception hall was filled with laughter

 Extended Version

and chatter as Jessica and Jimmy's family and friends gathered to celebrate their special day. The tables were laden with delicious food and drinks, and the atmosphere was joyful and festive.

As everyone sat down to eat, Mike stood up, his glass raised in a toast. "To Jessica and Jimmy," he said, his voice filled with emotion. "I've had Jimmy's back since he arrived in Germany, and I've seen the way he looks at you and how he talks about you, Jessica. It's a love that's truly special."

Jimmy and Jessica blushed but couldn't help smiling at each other.

Mike continued, "I remember the first time Jimmy saw you, Jessica. He had a look on his face that I'd never seen before. It was like he'd been struck by lightning. He was so stuck on you that all I could do was let y'all be."

# Replenishment

Jessica giggled, her eyes shining with happiness. "I remember that day too," she said. "I knew right then and there that Jimmy was someone special. I just didn't think that I would be stuck with him for the rest of my life."

Jimmy's face turned bright red, but he couldn't help grinning. "Oh, that's how you feel, honey. You love me, though, so I can live with that," he said, his voice filled with laughter. He looked at Mike, "I appreciate you, brother."

As Mike sat down, the music started, and Jimmy and Jessica took to the dance floor for their first dance as husband and wife. Jimmy's eyes locked onto Jessica's, his heart filled with love and adoration.

"I'm so nervous," Jimmy whispered, his voice trembling with emotion. "But at the same time, I'm so happy to be here with you."

 Extended Version

Jessica smiled, her eyes shining with love. "I'm happy to be here with you too," she replied. "I promise to always support you through all of life's ups and downs."

Jimmy's eyes filled with tears as he looked at his beautiful wife. "I promise to always love and cherish you," he said, his voice barely above a whisper.

As the couple danced, their love for each other was palpable, and their guests couldn't help but feel happy for them. Mike watched them with a smile, his eyes shining with pride. "I knew they were meant to be," he said to Philip, Jimmy's father. "They're perfect for each other."

Philip nodded; his eyes misty with tears. "They are," he replied. "They're a match made in heaven."

Extended Version

# Replenishment

As the night wore on, the dance floor filled with laughter and music. Jessica and Jimmy continued dancing, surrounded by their loved ones. It was a truly magical moment, one that they would treasure forever.

Despite the absence of Jimmy's mother, the wedding was a testament to the power of love and family. Jessica and Jimmy had created a community that was strong, supportive, and full of joy. As they danced under the stars, surrounded by their friends and family, they knew their love would last a lifetime.



Extended Version

# Chapter 5

Answers

Jimmy had just woken up and was sitting on the bed. Jessica had woken up extremely excited and ready to start her day. Jessica and Jimmy had been talking about trying to have children. They had an appointment set up so they could discuss their options, and Jimmy was going to get checked to see how healthy he was and if they would be able to have children naturally or if they would adopt. They had tried a few times on their own in the past, but Jessica never got pregnant.

Jessica and Jimmy sat down and had a conversation about seeking some professional help. Jimmy was very private, so Jessica had to do some convincing. Jimmy was a tough guy on the battlefield but quite a softy when it came to talking about personal problems. He was pretty embarrassed about not being able to get Jessica pregnant, so today would be a challenge for him. It was also a little depressing for him.

Jimmy got out of bed after noticing Jessica was up and moving around. He kissed Jessica on the forehead. "Good morning, honey. How did you sleep?" he said.

"I slept well. I'm excited about today," Jessica replied.

"Yeah, I can tell. You're up, bright and early. We still have a few hours before we must leave," Jimmy said.

"Yeah, I know. I'm just anxious; you know how, I guess," Jessica said.

"Yeah, I know," Jimmy replied, looking a little down. He wasn't as ready for the day as she was. "Well, I'm going to get up and get myself together."

 Extended Version

"OK. Well, you want me to make you some coffee. I'm also going to make something for breakfast so that we are good before we head out," Jessica said.

"OK, sounds good. Coffee sounds great. You already know how I like mine," Jimmy said.

"Yeah, I think I can manage," Jessica said.

Jimmy headed to the bathroom to shower and brush his teeth. After showering, he found something comfortable to wear and got ready for the day.

Jessica came into the room and placed Jimmy's cup of coffee on the nightstand. "Here you go, honey," she said.

"Thank you," Jimmy said. He grabbed his coffee, took a sip, grabbed his phone, and headed to the kitchen to see what his wife had prepared. He could smell the food all the way from the room, so

Extended Version

his mouth was watering; he couldn't wait to eat. Breakfast was his favorite meal of the day–the one thing he was looking forward to.

Jimmy didn't want to talk about his problems, and he surely didn't want to hear any bad news when it came to his manhood. Men are more self-conscious when it comes to personal issues than women. Jimmy hardly liked to go to the hospital because he didn't want to hear any bad news, so this time wouldn't be any different to him.

Jimmy finally made it to the kitchen and sat down at the table. Jessica came over and placed a plate in front of him. The plate had sunny-side-up eggs, country-smoked sausage, hash brown casserole, and wheat toast. Jessica also placed some grape jelly and a nice big cup of orange juice on the table; she made a meal fit for a king.

 Extended Version

# How Deep Does It Go

Jimmy smiled as his wife walked past. "The only thing that looks better than this food is the sight of you walking away," he said.

"Don't start; it's too early for all that," Jessica said.

"Says who?" Jimmy asked.

"Says me," Jessica replied

"I guess it's cool since I got this food to keep me occupied," Jimmy said.

"I bet," Jessica said with a smile.

Jimmy and Jessica finished their breakfast, gathered all their belongings, and headed out. Forty minutes later, they pulled up to the doctor's office.

"You ready, honey?" Jessica asked, looking over at Jimmy.

"Yes, as ready as I can be," Jimmy replied.

Extended Version

"OK," Jessica replied, even though she could tell he was a little bothered about something. "Thanks for coming," she added.

"You're welcome," Jimmy replied.

They got out of the car and headed into the building. The receptionist was ready to check them in as soon as they walked into the building.

"Do you guys have an appointment?" the receptionist asked.

Jessica looked over at the clock. "Yes, we are supposed to see the doctor at 10:00 a.m."

"Is this your first time here?" the receptionist asked.

Jessica looked at Jimmy, who replied, "Yes, it is."

"OK." The receptionist handed them a clipboard. "Fill this out, and we will get you checked in."

"OK," Jessica replied as she took the clipboard. She walked to find a place where she and Jimmy could sit.

Jessica took control of filling out the paperwork because she knew they would probably be waiting for a while if it was in Jimmy's hand. She could tell he didn't feel too comfortable being there. He had been on edge since he woke up that morning. His energy was way off, but Jessica paid no mind to it, as that could put a damper on her day. She was excited to know what their future held. Whether it was good or bad, she was prepared to make the best out of it and looked into all the options on the table from that moment forward.

"OK, well, I'll go over everything. OK …" Jessica said. She looked at the clipboard and started reviewing some of the paperwork information:

*Patient Instructions*

Extended Version

# Replenishment

*Semen Analysis Checklist • Semen Specimen May be collected at the Clark Med Center. If collected at another location, not at the Clark Med Center, please see transport instructions at right (#4). • Completed Physician's Order Form (Clark Med Form # SP-017) • Completed Collection Information section below. PLEASE NOTE: Instructions, locations, and hours are different for Fertility and Post-Vasectomy collection. Please see #4 at right and collection locations on back.*

*Semen Analysis Collection and Transport Instructions*

1 *REFRAIN from sexual intercourse and masturbation for a period of 2 to 7 days prior to collection of the semen specimen.*

 Extended Version

2  *Your physician or Clark Med Labs personnel will provide a sterile cup for collection. Print your NAME and DATE OF BIRTH on the cup.*

3  *Collect the complete sample by MASTURBATION in the container. Avoid the use of lubricants. Coitus interruptus (intercourse stopped before ejaculation), oral collection, and specimens collected in condoms are unacceptable. Write the TIME of collection on the container.*

4  *During transport, keep the sample protected from extreme temperatures and light. • Fertility: Deliver the sample WITHIN ONE (1) HOUR of collection directly to the Raleigh Medical Park Laboratory. Any delay will cause living sperm to die and give an inaccurate*

Extended Version 

*assessment of the semen. • Post-Vasectomy: Deliver the sample WITHIN TWO (2) HOURS of collection directly to the laboratory of choice.*

5    *Complete the collection information form below.*

6    *Submit your physician order form with billing information. (Clark Med Form #SP-017).*

7    *Results will be faxed to the referring physician who will contact you for discussion of the results.*

"Well, let's go one by one now. Just fill in the blank. OK. One," Jessica said.

"Hard as it was, we haven't had sex for eight days," Jimmy said.

"OK, all the numbers down to five are already complete. So, we just have to fill in our personal information and our billing information while you

    Extended Version

take the cup and go into the bathroom to provide the sample," Jessica said.

"OK, I'm on it," Jimmy replied.

Jimmy got up and headed to the counter to grab a cup from the receptionist so he could take care of his part. With a smile on his face, he glanced back at Jessica as she filled out the paperwork, just hoping that everything would go well, and they could have kids with each other. He wanted to be able to fulfill all her needs and more. She meant the world to him; it would be wonderful if he could give her the world.

# Chapter 6

The hunt begins


**Two months later**

Present day:

An 85 Ford pickup truck came to a halting stop off to the side of a dirt path close by the hunting ground in the woods of Colorado. Norman stepped out of the driver side and made a loud 'plop' as his boot hit the dirt road, stirring up dust as he planted his foot firmly on the ground. Norman was an older guy who enjoyed the open land and loved the hunt of the game. He had been in the area for a while, and he knew just where to find the big bucks. He wore his brown and green camouflage, with his boots. He had finally got and eased his head back to get a nice gulp of the forest scent.

                Extended Version

"Now Douglas, there is nothing like being out here in the woods–just us and the buck."

"You got that right. I'm going to get a nice buck today. You took home the prize winner last week, but it's a new day and I'm feeling lucky."

"Luck is what you're going to need because this rifle will bring down the biggest of 'em," Norman replied as he pulled his rifle out of the aluminum box that was on the back of his truck, "It's going to be a mighty fine day."

Douglas reached behind his seat and pulled out his lucky hat.

"Yes, it is, Norman. Let's get out of here and find a good spot so we can knock down a buck to take home for supper."

They both gripped their rifles and walked into the woods. A few yards into the woods they stroll. "Just up there we will post up there for a while and

Extended Version

wait for that big boy to come through here," Norman said, pointing at a tall bush, Norton had broken a couple of branches so he could climb in nice and snug. While walking over to the bush and Douglas replied, "I have a good feeling today."

Douglas pushed the brush to the side so that he could climb in.

This day was a little different than most. The forest was usually more active with birds chirping and rodense running the forest floor. There was usually a gentle breeze, but there wasn't a leaf or a bush swaying. It was like they were looking at a picture. As gorgeous as the scenery was, the forest just seemed as if it had no life. The forest was so quiet that the only sound they could hear was the clutching of their gun and the short impatient breaths that they were taking. The sun beamed through the cracks in the forest décor, settling on Douglas's forehead.

 Extended Version

"Man, isn't it hot today?"

"Shhhhh … I think I hear something."

The shadows consumed the left side of his face. His eyes locked in and his ears harvested every sound, as the broken branches crackled, as something crept across them gracefully. A few minutes had gone by, Norman slowly lifted his rifle so he could be ready to lock on when the buck was in view.  A few more minutes had gone by, the buck appeared and to Norman's amazement, this was the biggest buck by far.

"Wow, that is an award-winning buck. He is a beauty," Norman whispered, with his finger on the trigger. He took in a breath, looked through the scope, locked in and as the deer fed, he squeezed the trigger. POW!!! the rifle sounded off; it was the only sound that echoed through the forest–like a tree had crashed into the forest floor. The bullet flew toward

Extended Version

the buck and the fire from the rifle sparked in the buck's eye. Before the buck could reach, it cried out an awful sound as its body fell to the forest floor. The men climbed out of the bushes and stood over the buck.

"Wow, he's huge. I'm surprised that one bullet took him down. You really outdid yourself this time. I'm never going to catch up with you."

"A little friendly competition never hurt. You will have your time," Norman replied, as he looked at the buck. "What in the world!"

They looked down at the buck and its body started to transform into something that they had never seen before.

"Let's get outta here!" Douglas yelled. They started running, before they could get a yard away, a small black sphere flew at them, a light flashed and before they knew it their souls were being stripped from their bodies. Then the beam of light

 Extended Version

disintegrated their bodies into a withering nothing, vanishing them from the face of the earth. Leaving behind nothing but their weapons and a smoking pair of boots.

The sphere retracted backward into an Orange-reddish-scaled hand, which clinched shut after catching it then the creature ran off into the woods.

Meanwhile, Jimmy had finally made it home and his wife was so happy to have him back. Even though he was home, he still had to take it easy; being in a coma for months only made his body weak. He was to the point where he could stand up, but his legs were as wobbly as an elastic band. Every step that he took was as if he had walked a mile. He was to the point where he almost felt handicap. He was so used to doing things on his own; he was growing frustrated about the whole situation. Every time his wife ran to his rescue, it made him dread life itself to the point where he would just stop and sit on the floor wherever he was and tell his wife to let him be.

--A few months later–

Jimmy was able to move around a lot better as the months went by. He could walk around just fine now. His muscles were starting to work a lot better, but he was still fighting against his nightmares. No matter how much he tried, he couldn't stop the incident from playing over and over in his head. All he could see was the plane coming back around and as soon as the missile started coming toward him, he would come to and would be drenched in sweat, like he had been sitting in a steam room. His heart would race, and his eyes would be wide open, like he had woken up from a nightmare. Those events

 Extended Version

were taking a toll on his body and were compounding sleepless nights like stars in the galaxy.

Jessica woke up plenty of nights for the last few months that he's been out the hospital, comforting him as he went through the motions in his sleep. She would sit up on the bed and place his head on her lap as she rubbed her fingers across his head, trying to soothe him as if he was a disturbed child.

Even though Jimmy had made it back home, Jessica felt as if she still didn't have him back. She still felt she was alone at times–because, some days, he was himself and others, he was lost in pain. One morning, he woke up in a drenching sweat, jerking up on the bed as he sat there, looking around in a panicking state. His wife turned to him and sat up

Extended Version

on the bed as well. She rubbed his back as he flinched.

"Honey, I'm here. It's just a dream," she said. He looked over at her, eyes bugging.

"Honey, I think it's time that you see a therapist about this because it's not getting any better, only worse. It's been a few months now since you have been out of the hospital and you're still having those horrible dreams"

He took a deep breath. "Maybe you're right. I will make some calls later today and see when I can start some treatment because I don't know how much more of this I can take. I'm already in enough pain from this metal plate in my head; it's a lot better than it was, but I can't deal with both. One or the other has to go because it's driving me crazy." He looks at Jessica. "Thank you for sticking by my side through this. I know it hasn't been easy and I just want to thank you. I'm trying to fix myself, because

        Extended Version

you deserve so much better, and I want to be able to give that to you."

Meanwhile, at the police department, Phillip was at his desk and his partner Wayne walks up. "So, how's it going, partner?"

"It's going good; you ready to get to work? We got a couple of calls about two missing guys. A Norman Horn and a Douglas Fitzpatrick; their wives said they were going hunting together and they hadn't seen them since that day.  Mr. Justin Dean wants the case, but the captain has given it to us to see if we can locate them first. The missing persons' paperwork had been filed a few months ago, but no one has a lead, so I wanted to pick up this case because I feel like this might be tied up with the missing people's events that seem to happen every 25 years around the same time. Benjamin Plessner used to be obsessed with these events, but he ended up getting checked into a mental hospital after reporting that he had seen an alien, which vanished in the middle of nowhere while he had it cornered. I don't believe it's aliens, but I do believe that there may be a psychopath killer out-there somewhere. It's just crazy how he hasn't been caught yet after all these years."

"Aliens, ha-ha ... Wow! Yea, he really has lost it. So, how would you like to start?"

"Well, let's go back to basics and question the families and see if we can piece together something that might have been looked over. The little clues are the ones that solve the big cases, and we need to locate the truck that they were driving. If we can find it, we can possibly find some clues that we need to find them. So, put out an APB on the license plate and let's see if we can locate his truck," Phillip said, shuffling through some papers on his desk and he looked up and noticed Dean. After Dean got chosen

 Extended Version

for the Special Forces over him, things just haven't been the same. Phillip has a grudge against Justin. Justin eats it up every day they cross paths.

"OK, anything else?"

"Speaking of the devil, there's Dean," Phillip said, taking a deep breath as he shakes his head.

"Don't worry about him; keep doing what you're doing, and you may become the captain one day. Besides, you're the best detective that I've ever had the chance of working with," Wayne whispered. "Here he comes."

Dean walked up, grinning. "Hey, fellas, how is it going? Phillip, I heard that you are on that missing persons' case. How is that going for you?" Justin asked sarcastically.

"Do you mean the case that you couldn't get?" Phillip replied.

Wayne smirked.

Extended Version

# Replenishment

"Trust if I wanted it that bad, then I would have it. Just got another case I'm working on at the moment, but I have to roll. Peace," Dean said; he walked away with a straight face that remark really made his blood boil.

"Ya, I beat," Wayne mumbled.

"That guy is a real douche; I can't see how guys like him even make it in the academy. But anyways, go handle those things so we can get up out of here and solve this case," Phillip said, his desk phone started to ring. He waved Wayne on and picked up the phone. "Hello, this Detective Howell. How may I help you?"

"Hey, Dad, how's it going?" Jimmy asked.

"I'm great, son. How are you holding up? I know you have been going through it with the accident and surgery."

"It's been hard, but today, I'm going to find a therapist to see; that way, I can figure out how to get

 Extended Version

past this. Jessica has been hanging in there with me, but I need to get ahold of this thing."

"That sounds like a good idea, son. Well, you know, I will support you in anything that you do."

"Thanks, dad. So, how is mom?"

"She's OK, but she still hasn't gotten over that."

"She's still holding that against me. I'm going through a lot right now and it hurts that she didn't even come to the hospital," Jimmy said; his father could hear the hurt in his voice.

"I know she wanted to come, but her pride wouldn't let her."

"Pride, really? Well, I'll talk to you later, dad," Jimmy said, hanging up the phone, not even giving his father time to respond. The only thing his father heard was the dial tone and he just sat there for a second and then slowly lowered the phone, hanging it up.

Extended Version

# Replenishment

Phillip shook his head.

"I can't wait until they get over this."

His partner came back to his desk. "OK, I got that APB put out and I got the addresses for the wives, so we can take a visit."

"OK, well, let's get on the move," Phillip replied, he jumped up from his chair and grabbed his coat. Phillips and Wayne headed out the station. They had made it to their cruiser. "Let's do this," Phillip said; they hopped in the car and drove off.



Extended Version

# Chapter 7

Investigation

A short while after, they had made it to the Horns residence. They parked and got out the car and walked up to the front door. Phillip knocked on the door and waited patiently. Mrs. Horn came to the door, answering, "Who is it?" She looked through the peephole. She noticed it was the cops and her energy dropped because she knew that either they hadn't found her husband yet, or they had bad news.

"Yes, this is Detective Phillips," he replied as she opened the door.

"Hello, officers, how may I help you?" Mrs. Horn asked.

"Yes, I'm Detective Phillips and this is Detective Banks. Do you mind if we asked you a few questions regarding your husband's disappearance? We have been assigned to the case and we will make sure we figure out what happened to your husband," Phillips said.

Wayne nodded in agreement.

"That's the same thing the last two officers said, and it's been a few months, and he still hasn't been found, so I would rather not talk to anyone at the moment. I just want you people to find my husband," Mrs. Banks shouted in an aggravated tone.

"Ma'am, please wait; I know this is a hard time for you. We want your husband to be home just as much as you. But if you could just help us by answering a few more questions, it will be very helpful in locating your husband's whereabouts," Phillip said in a caring voice as he looked at her.

Her face told a million stories. No matter how much she wanted to slam the door in their faces, she



Extended Version

knew they probably were the only ones who would be able to find her husband.

"OK, come in," she replied, leading the detectives into the living room. "Have a seat," she said, waiting for the officers to sit. She took a seat across from them. "What would you like to know?"

Phillip looked around the house. He noticed a deer head hung over the couch. He also noticed pictures over the fireplace, with Mr. Horn holding a rifle while a buck lay at his feet. "So, your husband is a hunter? I see."

"Yes, he loves hunting. He and Douglas set out a few months ago to do a little hunting," she replied with a sad look on her face, almost not wanting to talk about it. She knew she had to, so she was making an exception just for today.

Extended Version 

# Replenishment

"So, Mrs. Horn, how close was he and Douglas?" Phillip asked, looking at her face for some type of unsettling expressions.

"He and Douglas have been friends for some time now. He met him about ten years ago while we were out of town on vacation. We were doing karaoke and he and his wife ended up doing a duet with us. After that night, they saw just how much they had in common. We had no idea but to find out that they lived a couple of blocks away. From that moment on, they started hanging with each other. They got drinks together, they barbequed together, and they loved to go hunting together," she replied with a smile on her face, which started to die down after a few seconds of thinking about the past.

"Did they have any fallouts before they went on their recent hunting trip?" Phillip asked Mrs. Horn.

 Extended Version

"No, not at all. They were as close as brothers would be," she replied, trying to make out what he was asking that question for.

"OK, just was trying to weed out a few things. It would be hard for anyone to kidnap or hurt two men with rifles. So, if they were that close, maybe they got stuck somewhere." Phillip continued to look around. A photo had caught his attention; the place looked somewhat familiar. "Do you mind if I look at a few pictures of your husband that he took while he was out hunting?" Phillip asked, having his eye on one picture in particular.

"Go ahead, Detective anything that may help. Be my guest," Mrs. Horn replied as she watched Phillip go straight for the picture that he had recently taken. He examined the picture quite thoroughly. "This place looks so familiar. How old is this Picture?"

Extended Version

# Replenishment

Phillip looked back at her, holding the picture in his hand.

"That is a fairly new picture. He took that at a new spot he found. He told me that it was his new spot and that he had seen some of the biggest bucks there. He was so determined to set a record for shooting the biggest buck, so he was always looking for the next big spot, and he had convinced me that he had finally found it."

"OK, I see that he had a nice-sized buck in this picture."

"Yes, it is. That's why I became a believer once he dragged the buck from the truck to the back yard."

"Well, he has defiantly got my vote," Phillip said, shaking his head at the astounding picture. "He defiantly has. Well, thank you for your time. I think we are going to head out of here now. You have certainly been helpful, and I recognize that this has

 Extended Version

been a tough time for you, and I promise I will do my best to find out what happened to your husband."

"Anytime, and thanks for stopping by. I'm happy that someone is still on the case to find out what happened to him," she said as she escorted the detectives to the door. "Be careful out there; something weird is going on out there; I can feel it. Something just isn't right," she said, waving to them as they walked down the driveway.

For some reason, she felt a little better about Detective Phillip; she felt that he would actually find out what happened to her husband. She didn't feel this way about any of the other detectives who came by and spoke to her. It was defiantly something different about this time.

"Bye, Mrs. Horn. I will contact you soon," Phillip said as he and Wayne approached the car.

# Replenishment

They both jumped in the car, honking the horn as they drove off.

Meanwhile, Jimmy had made it to the therapist's office and was signing in so he could see if he could start fixing the events going on in his head since the accident. Not only was it driving him crazy, but it was also causing a damper on his marriage. He felt bad that he couldn't be the man he used to be and was in a state that he was willing to do whatever it took to get back to that point.

He could look into his wife's eyes and see that his events were starting to break his wife down. She puts up this wall like she can handle it, but he has caught her crying a few times. That is one reason why he decided to see if he could get help to help him deal with his problems. He wanted something to ease his mind; he didn't care for medicine, but at this moment he was willing to try anything.

 Extended Version

# How Deep Does It Go

The way the war was playing back and forth in his head would certainly push him over the edge soon. One moment, he's in the war and things are pretty bad, and the next, he's waking up with a killer headache and things went from bad to horrible. Life has surely thrown him a curb ball, but he would figure out a way to hit it out the park.

"Thanks for signing in, sir. Miss Evans will be with you shortly," her assistant said. "Have a seat."

"OK, thank you," Jimmy replied. He took a seat a looked up, noticing the TV was on the news and people were starting to disappear all over. It was almost like a tragic event, for none of them had been found. There were even a couple of people missing in his hometown; they had a list of names, and some families were offering rewards if found. It was also said that this was something that has been happening every 25 years. 25 years ago, thousands

# Replenishment

of people disappeared within a few weeks. TV: "If you have any information pertaining to the whereabouts of any of these people, contact us immediately. 1-800-SAV-THEM."

"It's that time again; they are going to take who they want. It's like our planet is their farming ground. It happened 25 years ago, and it's happening again. They come when they are ready and take who they want," Benjamin said; he turned and looked at Jimmy right in the face. "Avoid the darkness; that's when they hunt," Benjamin said.

He was so shook-up from what the old man was talking about; he didn't notice Miss Evans had walk out. "Jimmy," she called out.

"Yes, I'm here."

Jimmy put his hand up so she wouldn't miss him. Standing to his feet, he walked toward her, looking back at the old man. He turned and reached out to shake her hand. "How are you?" Jimmy asked.

         Extended Version

"I'm good, thanks. Please come this way," she said, opening the door so he could walk in. "You can take a seat over there," Amy said, pointing at the couch. She walked to her desk and grabbed her binder and then walked over and took a seat in a chair across from him. "So, can you explain briefly what's going on with you?"

"Well, I keep having flashbacks of what happened when I was in the war. I always see a missile heading right toward me, and when I get pushed out the way, I wake up in a horrible sweat. It's getting more constant, and I can't sleep at all. It's not just keeping me up; it's keeping my wife up also. She wakes up when I do and holds me until I go back to sleep. She and I are both having a hard time with these nightmares. I just want them to stop," Jimmy said, getting a little emotional. Even though he was wide-awake, it still bothered him to talk about it.

# Replenishment

The dream wasn't only real; it was a reality that almost took his life. He was grateful to be alive, but if this were something that he had to continuously deal with, then he would have rather died that day.

"Calm down, Jimmy. It's just you and me here right now. Never let anything take you to the point where you want to lose control. Dreams are just figments of our imagination, which we make real–if we allow them to be real. Our dreams will take over our lives if we allow them too. What I am going to do is, give you exercises to try so we can see if you can gain control over your dreams.  Dreams usually come from something that we play over and over again while we are awake, and then they follow us when we lay down to sleep and start to play over and over while we are asleep. So, we are going to work on stopping that playback where it starts; then we will be able to stop it all together," Amy said, writing notes in her binder. "So, I am going to give

 Extended Version

you a couple of techniques to start off with. So, start off with these few things; then we will go from there," Amy said; she handed him a piece of paper.

"That's all?" Jimmy replied, looking at the piece of paper which only had a few things written down on it. "Are you sure this will work?" he asked as he looked at the few sentences.

"Trust me; so, go home and do this exercise with your wife. I will schedule you for a follow-up appointment on Thursday at 2 p.m."

Amy continued to take notes.

"OK, I will see you then; make sure you sign out at the desk. Can you close the door behind you? Thank you," Amy said.

He got up to exit the room. "OK, I'll see you Thursday," Jimmy said, closing the door behind him.

During intervening time, Detective Phillip had pulled up to the location that he had seen in the picture at the Horn's residents. He pulled up and took a second to recap the picture as he scoped the area. "Wayne, I think this is the place. This is the same place my dad used to bring me when I was small; that's why it seemed so familiar," Phillip said, looking around. "Come on; let's check it out so we can see if we can get any leads."

"OK cool," Wayne said, as he stepped out of the car. He looked down at the ground to see if he could notice any tire prints. "Umm," he said, squatting down, looking at the dirt road.

"What is it?" Phillip asked as we walked over closer to him.

"Well, I found some tire prints, but they look fresh, so this couldn't possibly be the truck that they were driving in."

Wayne looked around further.

"I'm not so sure about that. Since we got here, I haven't felt a breeze of any sort, which is kind of



Extended Version

weird to me. We are in the forest; there should be some type of breeze."

Phillip looked further down the road and started to follow the tracks to see if he could find the end. He and Wayne walked down the road, following the tracks; about 50 feet down the road, the tracks ended. "That's weird; the tracks just end, there are no signs of where the truck attempted to turn around. It's almost like it just vanished after it got to this point," Phillip said; he pointed to the area where the tracks ceased. "It would be impossible for the truck to back up and stay within the same lines all the way out," Phillip said, looking back down the road.

"This is crazy, but I do feel like we at least have somewhere to start," Wayne said with a look of confusion on his face. "I wonder what could have possibly happened."

Extended Version

# Replenishment

"I'm not sure if I really want to know," Phillip replied, looking into the woods that surrounded them. "We can start over there; it is a path just beyond that tree. It's not too far from the tire marks. Hopefully, we can get some sort of lead so we can at least have a direction to start heading," Phillip said, holding his hand close to his pistol, for he wasn't sure what was awaiting him and his partner.

"OK, let's go."

"Keep your eyes open," Phillip said, looking behind him to make sure no one was watching as they entered the woods.

They scanned the wood deliberately, looking for any track of a bullet, print, or anything that they could run with. They walked a little deeper into the woods and noticed a few broken branches, like someone was trying to go somewhere in a hurry. Phillip looked down at the ground surrounding the area and noticed two different sets of footprints.

 Extended Version

"Wayne, come check this out," Phillip said, bending down on one knee. "There are some footprints in this area."

"Do you think it could be their prints?" asked Wayne, looking down over his shoulder. "Looks like they were running in that direction," Wayne said, pointing in the direction of the footprints.

"Come on; let's check it out," Phillip said as he stood up and walked in the direction that Wayne was walking. "I'm sure if they were running, then they must have dropped something."

"What in the world in this?" Phillip asked; he approached a metal object that looked like it had been melted down. "This almost looks like a rifle that has been melted down, but it looks like the handle is gone … What in the world would a melted-down rifle be out in the woods in the middle of nowhere! It would have to be some extreme heat to

Extended Version 

melt something like that," Phillip said, reaching down to touch it. "That's weird; it's ice cold."

Before Phillip and Wayne knew it, there was commotion beside them–in the bushes. They pulled their weapons expeditiously, aiming right at the bush. They looked at each other; Phillip signaled Wayne to approach the bush. They eased over to the bush; not knowing something was lurking in the bush, which wasn't anything that they had ever dealt with.

There was an orange-red-head creature looking right at them as they approached the bush. The creature pulled out a small black sphere, clutching in as they grew closer. They held their pistols tight, and as soon as they eased forward to pull the bush back, the creature enabled a supernatural invisible skin, which made it invisible to the human eye. Quickly aiming his gun, Wayne



Extended Version

pulled the bush to the side. Phillip had his gun aimed right at the creature's face and didn't know it.

He lowered his weapon and looked behind the bush, moving his head to the left, and the back to the right and he and the creature's faces were a few inches away from each other. He looked straight, looking at nothing, but little did he know the creature was looking him right in the eyes. He slowly pulled his head back, and so did the unknown. "I guess it was nothing," Phillip said. "Hmm," Phillip muttered, looking back at Wayne as he let the bush go, and it swung back to its original state.

"Something just doesn't feel right out here," Wayne said; he looked back at the bush as they walked away. "Let's just get the metal and get out of here."

"Ya let's do that. Maybe we can get some prints off it, even though it's melted down," Phillip replied, pulling a rag from his back pocket. "Some of it didn't completely melt down, so we might be able to get a print or two," he added, grabbing the melted metal.

They both exited the wood, keeping an eye out for any unanticipated ventures that could occur on their way out. They were already startled from a bush that shuffled previously, which seemed to have moved all on its own. Well, at least that was the impression that they had. The weirdest things happen at the most unexpected times, and that was surely one event that they would not forget, for it had made both of their hearts stop in mid-beat.

Nevertheless, that was the least of their worries; the phenomenon that lurked in the bushes would take the world by surprise. 25 years had passed, and things were about to take a turn. May God so help them, for they have no idea what they are about to encounter. There are powers beyond anything that we, humans, have the knowledge to

 Extended Version

concur up, let alone defend against. Replenishment is about to take effect, whether we want it to or not. It could be the universe taking its toll or perhaps survival enhancing itself. Only God knows and all we can do is pray that He is on our side with this one. Phillip and Wayne loaded up and headed out.

# Chapter 8

The unexpected

Later, that night, Tod had just gotten off work; he had a cold, and he was sneezing uncontrollably. He walked out the door into the parking lot, and his car was all the way in the back of the lot all by itself. Usually, during the night, there would be small gusts of wind, but tonight wasn't like any other night. He sensed that something wasn't right. He looked up at the sky, and there wasn't a cloud in sight, but the stars light up the sky as if they were shining just for him. It was tremendously quite, so every step that he took clunked on the pavement as if he were walking on a tile floor. Every step he took made him ever so more nervous. He felt as if

 Extended Version

he were being watched, so he stopped in his tracks and turned around, scouting the sense.

He had his coat in his arms, squeezing it tight, and he sped up his walk toward the car. Suddenly, he heard a vibrant whirl of wind, but he felt nothing, nor did he notice any trees or power lines swaying. He stopped and looked around once more and before he could cover his face, a sneeze thrust up his throat cavity and created a mist of germs that balled up in the air like smog and as the light hit the particles of snot, Tod could see a face figure being formed right in front of him.

Before he knew it, time seemed as if it slowed down. His heart stopped, and his body shivered in fear. There was an invisible being standing right in front of him and he felt as if he wanted to shit himself. He turned around and sprinted toward his car like a heat seeker missile.

Extended Version

# Replenishment

He made it to his car and was fidgeting to get his keys out of his bag, and before he knew it, a black sphere was hovering in front of his face. He froze in his tracks; a bright white light illuminated around the sphere, blinding him. Before he knew it, his soul was being sucked from his body, his body disintegrating into a miserable nothing as his white spirit got pulled into the black sphere. His clothes dropped to the floor, and the creature held out its hand, calling the sphere back, and then it pulled out a weapon of some sort and aimed it at the car. It scanned the car and clothing that were spread on the floor-just by the car-and then it pressed down on a red button that was simultaneously flashing on the side of the weapon, causing the vehicle and clothing to vanish into mid-air.

Back at Jimmy's house, he was sitting in bed, having a conversation with his wife about his therapist visit. He was a little overwhelmed with

          Extended Version

what the mysterious man had told him in the waiting area of the therapist office.

"While I was waiting, something came on the news about people going missing and there was this strange guy who started talking about how they were back and that they come every 25 years. He said it was like our planet was their farming ground."

"Whose farming ground?"

"He didn't really say who; all he said was they. I didn't really have any time to talk to him. I wanted to ask him what he was talking about, but as soon as he finished talking, my therapist called me. He was acting really strange; he had a look on his face as if he were trying to warn me or something."

"Warn you about what? You do realize that he was in a therapist's office, so he could really have something seriously wrong."

# Replenishment

"Ya, I know, but it just felt different. He also told me to stay away from the dark because that's when they hunt. I'm not sure what that meant, but it really sounded creepy."

"Honey, I'm sure it's probably nothing."

"I hear you, honey, but it's just crazy that people are missing, and this guy was talking like he knew why," Jimmy said; he looked out the window into the darkness. "What if there is something out there that's taking people?"

"I know it's crazy, but can you leave it alone? Because you're weirding me out a little. We are out here in the middle of nowhere."

Looking over at his wife, Jimmy said, "Sorry, honey, I didn't mean to scare you." He reached over and hugged his wife as they sat up on the bed. "I won't let anything happen to you."

"Yes, I know, but ghost stories aren't helping," Jessica said, smiling.

 Extended Version

Jimmy laughed. "Sorry, honey, I didn't mean to scare you." Jimmy looked out the window, wondering to himself, *I wonder if something is really out there.*

The darkness pierced his eyeballs like the sun would on a bright day.  He didn't really know what it was that he was wondering if it existed, but whatever it was, it got the man to believe. For some reason, the story felt so real; it didn't feel like a story the man was making up. He had never been so intrigued in his life about anything, and this surely had his attention. He wanted to let it go and he wanted to know more as well. He wanted to know what the man was referring to when he said 'they.'

The though seduced his mind throughout the night so nothing else could take vogue in his mind for the time being.  Even though Jimmy had once again begun talking to his wife about what the

Extended Version 

therapist had told him, all that was really on his mind was this unexplainable entity that was here on our planet, picking us off one at a time. The thoughts clouded his mind until he lay down and fell into a deep sleep.

Later on, that night, it was about 3 a.m., and something was scrambling outside their bedroom window. Suddenly, a head slowly crept up and it starred into the bedroom window and the creature planted a scene in Jimmy's mind, right into his dream state.

***Dreaming***

Jimmy jumped up out of his sleep from a cold sweat; as he sat on the bed, he heard a gust of wind, even though not a tree was swaying outside of his window when he glanced over. He got up and walked closer to the window so he could look out.

 Extended Version

As soon as he went to look out the window, he heard footsteps running across the floor. He quickly ran over and grabbed his pistol from under his side of the bed, raising the gun up and aiming it at the door as he eased over toward it.

He reached down and slowly opened up the door, with his gun aiming for a headshot. He slowly swung the door open as he stepped back, placing his other hand on the gun as well. Nothing was there, so he casually moved out of the room, looking to the right, and then to the left and back to the right quickly. Stepping out the room, he headed down the stairs, creeping slowly, one step at a time, as he scanned the room on the way down.

He noticed that the door was open, so he proceeded to the door, turning his body, with his gun gripped tight. It was so dark that he couldn't see anything. As soon as he reached to grab his flashlight

Extended Version

out of the drawer of the table by the door, a black shadow flashed by in the next room across from him. His heart started beating hard and a drop a sweat slide down the side of his face. The walls seemed like they were closing in, as he grew closer to the entrance. The darkness consumed the room like land does the earth, bringing all of his inner fears to life.

Suddenly, a flash of light lit up around the corner, and he hurried over, taking a deep breath while he braced himself before he jumped around the corner, with his weapon's safety far from on. Thrusting him-self around the corner, before he aimed, the flash was gone, and darkness set in once again. He quickly hurried around the corner, with his flashlight and gun aimed in the direction of the light.

Surprisingly, there was nothing there; he waved the light up and down at the wall and nothing



Extended Version

was there. He slowly put down his gun and turned around and as soon as he lifted his head, there was a dark creature standing right in front of him. He jumped back and fell over the table, dropping his weapon and when he looked up, a bright flash of light burst in his face, causing him to cover his face.

***Dream Ended***

Jimmy jumped up out of his dream state, drenched in sweat, frightened out of his mind. He jumped up as though he had just gotten out of the hospital. His wife was shaken up out of her sleep, and she got up and braced him as he shivered with fear.

"It's OK, honey. It was just a dream. I'm here," Jessica said, gently rubbing his back. "Calm down, honey."

# Replenishment

Jessica continued to hold him, rocking him gently. Little did they know that the creature was outside their window, easing back off into the night; the creature was gone. The images in Jimmy's mind seemed so real; all he could do was look around in shock;  he stared at the dark window seal.



Extended Version

# Chapter 9

Black sphere


The following day, Detective Phillip and his partner Wayne went back out the wood because they had a hunch that there was something they could use out there that would help them solve this mystery. When they pulled up, Phillip noticed that the trees were swaying, and the birds were chirping, and it all sounded naturally. It was totally different from the previous day. Things seemed a lot more normal.

They were ready to scout the area a little more precisely. Phillip had a hunch, and he was determined to find something. Both detectives proceeded back into the woods where they had exited previously. Even though the day seemed better, they still entered with caution. They both had one hand over their weapons.

"I'm sure we are going to find something today," Phillip said as he walked deeper into the forest. "This is the place where we found that melted metal," he added, looking around, kneeling on one knee.

Wayne looked ahead and noticed that the wind was blowing, and all the trees were moving- all except the ones that were about 50 feet in front of them. "That's weird," Wayne said, walking closer to the area.

Wayne had walked about 25 feet away from his partner, heading straight toward the area where the wind ceased to exist. Phillip had gotten up from his knee and started walking, trying to catch up with his partner. Wayne was a few inches from crossing an invisible barrier, and he was unaware of it. As soon as he raised his leg and his foot penetrated this

 Extended Version

barrier, an egg-shaped metal object was triggered, and a weapon was activated and twirled around the egg-shaped sphere, aiming right at Wayne.

Wayne took another step, and he was directly behind a tree. The sphere started to power up and shot a gold laser beam of light directly at the tree that Wayne stood behind, shooting right through the tree and Wayne's flesh. The beam of light was so powerful that it went through several other trees after it penetrated Wayne's body. Phillip heard a slicing sound, and when he looked up, he could see the light shattering through the tree and then through his partner and through the other trees. He pulled his weapon and ran toward his partner Wayne as he dropped to his knees.

"Wayne!" Phillip yelled. He ran toward his partner. Phillip looked to the left and he could see this egg-shaped object fading away until it was

Extended Version

invisible. He looked with confusion written all over his face.

He didn't know what this object was, so he took caution. He slowly walked toward his partner and when he got closer, he could see his partner gasping for air. "Partner, are you OK? What in the world is going on?" Phillip said, furious and scared at the same time.

Phillip tried to stay his distance because he didn't want to get fired at too. He eased over toward Wayne and kept his eye in the area where he had seen the egg-shaped sphere. As soon as he crossed the invisible barrier, the sphere reappeared and started to power up again, so Phillip Jumped back out of the way and he looked up and noticed the sphere disappearing again.

*What in the hell is that thing?*

The triggering of the sphere also triggered a band that was wrapped around the creature's wrist. The creature looked down at its wrist, and then scrambled back to its transportation capsule. The capsule was in defense mode, equipped with deadly weapons to defend itself. It's made of an ore element called descranium, which is stronger than any metal the earth can produce. The weapons that were attached to the capsule closed down into the capsule to make it completely smooth. The power is 100 times stronger than any human-made weapon. It shoots pulse gold beams of light so it would take care of itself, but the creature had to make sure its transportation device was not damaged.

Meanwhile, Phillip had reached over and snatched his partner from behind the tree and out of the line of fire of the sphere device. He didn't have a radio; for he was a detective, and his attire didn't come with a radio anymore, so he was unable to call for backup or, in his partner's case, help. He wrapped his arms under his partner's shoulders and started to drag him toward the car.

The more he pulled, the more he realized just how hard it was to pull dead weight. His partner was out of it, so this was a task that he wasn't prepared for this morning. No matter how much he wanted to give up, he didn't because his partner wasn't dead yet, so he kept dragging his body. He had dragged him all the way to the entrance where they had come in.

The creature had made it back to the capsule and scanned the scene to see if its ship was distorted in any way. Noticing that his ship was fine, it then scanned the 50-foot perimeter. Noticing a tree that had been damaged by the capsule, it crept over to the area and looked behind the tree and noticed a puddle of blood and a trail of blood leading to the east.



Extended Version

Phillip had managed to get his partner back to the car. He propped his partner up by the back tire and was able to radio for help. "Officer down, officer down. We are off Hump free and Wayne's down toward the end of the dirt road. Please hurry," Phillip said, still in a state of shock.

After he radioed for help, he decided to place his partner in the back seat. He had the back door open and was able to pick up his partner and lay him across the back seat. He then went around to the other side of the car and opened the other door and reached in, grabbing his partner under his armpits and then dragged him further into the back seat of the car. He wanted to drive off, but he had already radioed for help, so he decided to wait until an ambulance had arrived.

Phillip didn't realize that it would have been in his best interest to leave while he could; the

Extended Version

creature almost made it to their location and was approaching with extreme prejudice. Phillip was comforting his partner the best he knew how. He let down the windows and, in the back, on the floor, he found an old shirt, which he used to prop up his head.

"Please hold on, partner, help is on the way. We have been through too much for you to give up on me now," Phillip said; he looked at his partner lying there in agonizing pain.

Suddenly, the bushed busted open, like something flew right through them. The commotion caused Phillip to grab his weapon, jumping to his feet, aiming right in the direction of the commotion. He looked, but nothing was there. It reminded him of the previous day when he saw the bushes move, but there was nothing there.

After he had seen the sphere-shaped object vanish, he was extremely cautious. He held his

     Extended Version

weapon up, his hand trembling in fear. He could hear a breeze, but everything was completely still– even Phillip's heart almost entirely stopped in his chest. He had grown completely paranoid and was all over the place, looking around at everything.

He heard a gust of wind, but as he looked around, not even a leaf budged. Suddenly, as he looked, he could see a small black ball hovering in mid-air. He aimed his gun right at it; the ball was spinning tremendously fast and before he knew it, the ball had vanished. Being so scared, before he realized it, he had shot his gun and it hit the creature,

"Rooahh!" the creature moaned.

The bullet canceled out the creature's invisible shield and Phillip was able to see the alien being for just a second before it vanished again, running off into the woods. Phillip didn't know what do, so he

Extended Version

ducked down behind the door, not sure of what he had just seen. "Oh, my fucking God! Was that a fucking alien," Phillip said, scared out of his mind.

He was holding onto his gun, and his hand was shaking so bad that he had to grab his wrist with his other hand. In the car, he could hear a buzzing sound, and when he looked over, he could see that black ball hovering over his partner, so he raised his gun slowly to try to shoot the ball and before he got a chance, the ball released a bright light, blinding Phillip, causing him to put his arm over his face while it extracted his partner's soul from his body, disintegrating his remains.

The sphere flew away with extreme prejudice, and about time Phillip jumped up to shoot, he looked, and his partner was gone. All that was there was his clothing. "Wayne!" Phillip cried. He looked around and swung his gun. Phillip was ecstatic; he had never witnessed anything like this in his life.

 Extended Version

The only reason he was trying to hold it together was for his partner's sake.

"Wayne, where are you?" Phillip asked. He scanned every square inch around the car. "Wayne!" Phillip repeated softly. He eased around the car, watching his back.

He heard a breeze sound again and he turned and (STOP) fired his weapon out of fright; the bullet ricocheted off a tree. He heard the sound of breeze again, and it was like it was right behind him; he frantically turned and fired again, hitting nothing. He heard the buzzing sound again behind him, and as soon as he turned around, cop cars started screaming down the road. Phillip whipped around and before he could get his head all the way around, the sphere had disappeared and all that was in Phillip's line of sight was the cop cars and light that lit up the forest like Christmas lights.

Extended Version 

# Replenishment

A gust of wind ran across his face like a drought that had just ended. Phillip noticed that whenever the aliens were around, the elements seemed as if they froze in time. The trees swayed from left to right and the birds were chirping. It was like the forest had come back to life.

This was one of the weirdest events that Phillip had ever witnessed. He didn't know if he should tell someone what he just saw or keep it to himself. His partner vanishing was going to be a tragic event that he wouldn't be able to explain. Not only was he the last person to see him, but Wayne's blood also was all over the back of Phillip's car. That alone would make him a suspect.

He's been doing this long enough to know that they were about to take him into custody; not only was the blood in his car, but it was all over his hands and a trail of blood was leading directly to his back seat. The situation had gotten serious really quick,

 Extended Version

and Phillip didn't know a way out. He didn't want to run off because all fingers would defiantly point right at him, but if he didn't, he would be taken into custody and held until proven guilty. Since the alien wouldn't be caught, he would be held responsible for the disappearance of his partner. As much as Phillip didn't want to leave the scene, he really didn't have too many other options.

Right then and there, Benjamin's name popped into his mind like the answer to a question buried so deep in thought. Maybe if he were able to reach out to Benjamin, he would be able to find the answer to some of these weird events that had arisen. He knew that Justin was probably headed out to the scene, so if he needed to do something, then he had to make that choice now.

Benjamin was a retired detective, and he could remember how he used to talk about black balls and creator from another planet. Everyone accused him of being crazy, but after Phillip had seen some of the things that he just saw, he wasn't so sure if Benjamin was as crazy as people accused him of being.

Before the police could get close enough to see Phillip, he had fled the scene. Having so many questions that he needed answered, his duty was to locate Benjamin at any cost. Within the next hour, his phone would most likely be tracked, so he had to get all numbers that he may need later.

Phillip was a detective, so he always kept a pad and pen just in case he needed to write down a brief description of anything. After Phillip had gotten far away enough, he scanned through his phone and wrote down all his important numbers, including his son's. His son would believe anything he told him, for he had always been there for Jimmy-who talked to him about any and everything. They had a pretty strong bond, so he knew if no one else



Extended Version

believed him and then jimmy would. At this moment, he was his best hope of helping him find out what was going on.

Police cars were all over the scene, collecting evidence and trying to figure out what went down this morning on the old hunting grounds. Forensics had the area taped off and was getting prints from Phillip's cop car. They knew the car belong to Detective Phillip's, but he was nowhere to be found, and neither was his partner. So, they were trying to gather prints to see if anyone else was on the scene. If not, then a missing person's complaint would have to be filed with the two officers.

Dean pulled up and jumped out of his black expedition truck. The dust from the dirt road made a cloud of dust as he walked toward the scene. He walked right through the cloud of dust up toward the leader of the forensic team so he could try to get some answers as to what was going on. Justin was a country boy, so he was into the whole cowboy boots and tight jeans swag. He was a little cocky, but no one could tell him that. Although he was a country boy, his accent was nothing like it at all. His parents were well off, so this act he portrayed was only fooling his co-workers. Everyone who truly knew Justin knew that he wasn't what he put off. He only went out to become a member of the Special Forces because his father liked it.

"Can someone fill me in on what happened out here?" Justin asked, biting down on a toothpick that was hanging from his bottom lip like a Newport. He and the head of forensics walked away from everyone.

"I'm not really sure yet. We found a path of blood; we also found some empty bullet shells. It looks as if the detectives were shooting at something; I'm still unsure of what that was. It could



Extended Version

have been at each other, or it could have been at someone else. It's still hard to tell. We are pulling prints to see if we can place anyone else here at the scene, but if not, we will have to assume that one of the detectives was shooting at the other. The only thing that is weird is that there are only bullet shells by the car and none in the direction of fire," said the forensics chief.

"So, in other words, no one was firing back? I still don't understand. If they weren't firing at each other, then why is there a path of blood? It looks as if someone had been dragged from the forest to Phillip's car, and for that to have happened, one of the detectives would have had to shoot at the other in the wood and someone would have gotten hit; then the other officer would have had to drag the other officer back to the car. But that wouldn't explain who was shooting at whom from the car.

# Replenishment

There must have been someone else involved," Dean said, parading the scene, looking around. "Get that blood test done and get me those prints so we can have a better idea of what we are working with."

"OK, will do. I will have them to you as soon as possible," the Chief of forensics said.

Dean stood there, looking around at the forest. He knew something was up, but he just didn't know what it was. He took the toothpick from his mouth and flicked it into the bushes and went and jumped in the truck. He picked up the radio. "I need all units to keep a look out for Detective Phillip and Detective Wayne. If anyone hears or sees them anywhere, bring them in and contact me immediately," Dean said, placing the radio down and glancing at the woods on the way out. He didn't know what to think, with all people that were vanishing on a day-to-day basis. He knew there was someone or something

 Extended Version

else there; he just didn't know who or what it was.

# Chapter 10

On the run


$P$hillip had made it out of the woods and was trying to see if he could locate a place where he could use a computer so he could try to get the address where Benjamin resided. He wanted some answers, and he didn't know anyone else he could run to but him. He could remember the things he used to say about people disappearing and how they were abducted by aliens. Everyone thought he was crazy, and now Phillip wasn't so sure of that because he had just seen an alien today. He either had seen an alien, or he was going crazy. The town wasn't too far away, so he started walking.


 Extended Version

Meanwhile, Jimmy was up, facing the fact that he had seen aliens in his dreams that seemed so real. If he hadn't woken up from his dream state that second time, he probably wouldn't have been convinced. His mind had been so occupied with what was going on with his dream that he didn't realize that he wasn't having the flashback from the war anymore. His mind had switched over to the situation at hand–the aliens and if they really existed.

He wanted to talk to his father about the situation, but he didn't want to seem as if he was crazy. He would just start by telling his father about what the guy had told him and go from there. If his father thought it was a load of crap, he wouldn't hesitate to be honest and tell him. Jimmy had grown obsessed with this mystery. He wanted to really find out if it had anything to do with all the people disappearing all over.

Jimmy had finally picked up the phone to try to call his father down at the station. After he dialed the number, he waited as the phone rang over and over again without an answer. He knew his father was usually there to answer the phone, so he tried once more. He finally got an answer, but it wasn't who he expected it would be.

"Hello, this is Justin. How may I help you?" Justin asked.

"Hello, I was calling to speak to my father. Is he available?" Jimmy asked with a look of confusion on his face, wondering why someone else was answering his father's personal line.

"Sorry, he's not here at the moment, but as soon as we hear anything from him, we will let him know that you called," Justin replied.



Extended Version

# How Deep Does It Go

"What do you mean when you hear from him? No one has ever answered his phone. What's going on?" Jimmy asked clutching the phone, hoping his father was OK.

"I'm not at liberty to speak about the issue that is at hand at the moment."

"Issue at hand? Is my father, OK? Tell me something; I'm his son, and I have the right to know anything. So, you can tell me something," Jimmy said furiously. "You know what? Don't worry about it," Jimmy said, slamming the phone down.

After hanging up in Justin's face, he immediately called his father's cell phone. All that happened was, his father's cell phone kept going to voicemail over and over again. He tried several times and still got the same results. After the fifth time, he really started to get worried. When the voicemail came on that final time, he decided to leave a message. "Beep …"

"Dad, what's going on? I called your job, and I called you several times on your cell, and I got no response. I'm getting a little worried. This is not like you at all; can you please call me back when you get this, Dad? I just need to know that you're OK, with all these people disappearing all over. Please call me back, Dad," Jimmy said, hanging up the phone and sitting there for a few minutes.

Jimmy looked at the phone, hoping that his father would call him right back, but those minutes turned into hours, and after a while, every text or call that came through made him rush to the phone, looking and hoping that it was his father calling back, but it wasn't. After a while, he just decided to put on some clothes and go to his father's house and see if he was at home.



Extended Version

## How Deep Does It Go

In the meanwhile, Phillip had made it back to town, and he had a few dollars of liquid cash on him, so he went into the clothing store and bought a new shirt, some shades, and a hat to try and change up his appearance. After leaving the clothing store, he looked around, making sure no police were in sight. He pulled his hat down snugly over his head, so it was partially covering his face. He then proceeded to the library so he could get on the computer and try to locate Benjamin's address. When he got to the library, he had to apply for a library card because he had never been to the library a day in his life.

"Excuse me, Miss, do I have to apply for a library card in order to use the computer?" Phillip asked, looking nervous.

"Yes, sir," the librarian said, looking at him. "Are you OK, sir?" she asked, handing him the forms to fill out.

# Replenishment

"Yes, I'm fine," Phillip, replied grabbing the papers from her.

When he grabbed the paper, she noticed how shaky his hand was. She looked up at him, and the way he had the shades and hat on made her a little uncomfortable. After he had finished the papers and headed over to the computers, she called 911. "Hello, I would like to report a suspicious man. He's wearing a cap and shades, and he's very jittery."

"OK, what's your name?"

"I'm Miss Porter."

"Where is your location?"

"I'm at 4209 N Tuttle."

"OK, we will have someone come check it out. He's still there, right?"

"Yes, he is."

"Do you by any chance have a name?"

"Give me a second let me check to see what his name is on his paperwork."

     Extended Version

Meanwhile, it was almost 12noon, and the day was growing shorter by the minute. Phillip had finally sat down at the computer and started to search for Benjamin Plessner. He had created a blog online talking about people being abducted by beings from another planet. He clicked on the link to read up a little about it and see if he could get a little information on what happened earlier this morning. He was still in shock that his partner had just vanished in the middle of nowhere.

When he clicked on the link, all sorts of pictures of what these creatures looked like were on the site. They almost looked like humans without a mouth. They had similar eyes, and their heads looked just like ours. They had no facial hair anywhere on their bodies. Their hands and feet looked like ours as well. The only difference between them and us was their lack of hair, they had

Extended Version

no mouth, and their skin was an orange-red color. They almost looked as if they were like some type of military aliens. They had bands on their wrist and belts around their waist and around their right leg to carry weapons of some sort. "What in the world are these things, and why are they here abducting us?" Phillip asked himself as he continued to scroll through the blog.

While he continued to scroll, Miss Porter was finishing up her call with the police.

"His name is Jimmy Howell."

"OK, thank you, Miss Porter. Someone is on the way."

All calls that had anything to do with Phillip would be routed to Justin.

The radio in Justin's Truck went off, "There is a complaint of a suspicious man in a library at 4209 N Tuttle, and the suspect's name is Jimmy Howell."

     Extended Version

## How Deep Does It Go

*Jimmy Howell? That name sounds so familiar … wait a minute that's Phillip's son, but why would he be at the library?* Justin thought, biting down on a toothpick. "Hold up; that probably means Phillip uses his son's name so he can get some information," Justin said, picking up the radio so he could alert all units. "All units in the area, head to 4209 N Tuttle; it's Phillip," Justin said, busting a U-turn right in the middle of the intersection. His tire gripped the pavement like a car coming around the corner in the Daytona. The smoke bathed the sky in a cloud of gray mist. He flipped on his sirens, and the lights illuminated across the hoods of the cars that had avoided his unannounced turn.

Back at the library, Phillip was still scrolling through the blog; he saw a blog titled *when they are around*. It had information so people would know when the aliens were nearby. It read: *The elements would stop like time had frozen. The wind would cease to blow, the animal would go completely silent, and you would hear a gust of wind, even though not a tree in sight would shiver or sway. Not even a leaf would budge; it was like time had stopped, and you were the only one who was able to move freely—like you had become their prey—and they didn't want to be distracted in any way while they were getting ready to strike.*

He also found a section titled *their weapons of choice.* He had drawn out pictures of the different weapons. Under each picture was a brief description of what that weapon did. There was a picture of a gun-shaped weapon called the '*Retention Equalizer,*' and its description read: *A weapon that shoots a triple short beam burst of silver light. They use this weapon as some kind of memory eraser that causes a person to forget what happened the day before.* He also saw the picture of an alien's waist with a black band around it, labeled *Flexure Truss.* Under that picture was the description: *It was some kind of band that lit up and caused material matter to*

     Extended Version

*bend, creating a portal the alien could walk through to escape.*
Phillip had posted all types of information on this blog. He wanted to find something explaining the ball he saw hovering, so he continued to search, and he finally saw the black ball that burst into a ball of light, and it's called *Deaddimizer*, with the description: *A ball that pulls the soul from a human body, then breaks down the body into small particles the size of a grain of sand.* Phillip also went on to add: *The ball also has the ability to put whatever that was broken down back into its original state*; he called it a *Readdimizer.*

Phillip looked up and noticed the librarian kept glancing at him like something was up. Every time he would look up, she would quickly turn her head. He had a bad feeling, so he hurried through the blog. He didn't want to be at the library any longer than he had to be. He continued.

Extended Version

# Replenishment

"So, is he saying that this alien creature has taken my partner's soul?" Phillip muttered to himself; he continued scrolling down the page. He also saw pictures of these creatures looking as if they were transforming into animal shapes. "So, not only can they disappear, but they can also transform into animals. This is some stuff you wouldn't read about in a Syfy novel."

Phillip had finally found Benjamin's address. Phillip thought to himself, *the way this guy has all this information, one would think he was an alien annalist. He seems to know a lot about these creatures. I have to find this guy, though, so I can see if he has information on how I can get my partner back so I can clear my name.*

Phillip googled the address and saw exactly where Benjamin was located. Good thing it wasn't too far from where he currently was. Well, it wasn't if you were in a car. But on feet, it would be at least

     Extended Version

a 20-minute walk. Phillip had left the building and was starting to head to Benjamin's house.

He was headed down the sidewalk when he noticed a squad car coming right toward him. Even with his hat and shades, they still might notice him. He saw a newspaper on a bus stop bench about 10 yards away, so he walked quickly toward it, but the closer he got to the bench, the closer the squad car seemed to get. As soon as he walked up to the bench, the squad car was just about to pass by him when he grabbed the newspaper and sat down, putting the paper up to his face like he was reading up. The cop car passed by, sirens screaming, and the motor was roaring like they were trying to get somewhere fast. Phillip thought to himself, *that was a close call, and I have to be a little more careful, or I wouldn't be able to figure out how I can help my partner.*

# Replenishment

Phillip heard the tires of a cop car come to a screeching stop, followed by other cop cars, which arrived on the scene around the same time. Right in front of the library, Phillip knew something was up. Now he knew he would have to stay off the main streets because one slipup could cost him everything as well as his life in jail. Immediately, after he had put down the paper, he got up to walk away and noticed Justin pulling up in his truck. Phillip turned and walked away as Justin and the other officers surrounded the building. Phillip turned to the corner and was out of sight before they realized that he was no longer in the building.

                        Extended Version

# Chapter 11

Frazzled

Shortly after, Jimmy had made it to his father's house, which was being staked out by police in an undercover cop car. Jimmy paid them no mind; he was focused on finding his father at the moment. He walked up to the house, canvassing every window to see if he could see any lights or movement inside. Not a light from a television set glided around the window seal, nor were there any shadows left behind. He knocked on the door, and the sound echoed through the house like an empty bathroom covered in dusty old tile. The cops watched from a distance, waiting for a curtain to sway or the front door to creep slightly outwards, followed by a suspicious eye. But there was no one there to spark their intuition. Jimmy was left standing, then knocking repeatedly, getting the same disappointing results. He finally had come to the reality that his father wasn't home; he turned and saw the reflection from the binoculars hazing around the lens in the undercover cop car. Things just became real to him that something was truly wrong. His father wasn't at the office, and now the cops were staking out his place of residence.

"What has my dad done?" Jimmy whispered to himself, looking at the cop car as he got back into his vehicle.

Meanwhile, back at the library, Justin talked to the librarian. "So, Ma, what did Jimmy come into the library to do?" he asked.

"Well, all he wanted to do was use the computer."

 Extended Version

"Can you show me which computer he was on?"

"Sure," she said, getting up from behind her desk and walking over to the computer he was on. "It was this one right here," she said, pointing at the computer.

"OK, thank you, Ma," Justin said, flagging down the forensics so he could get prints done. "Can you run prints on this keyboard and get them back to me as soon as possible?" Justin asked.

"Sure, sir," replied the forensics.

"Just give me a second before you get started," Justin said, pulling out a pen. "So, what are you looking for?" he asked himself, hitting the space bar with his pen. Benjamin's address popped up on the screen. "So, what do you want with crazy Benjamin?" He walked over to an officer. "Get a couple of guys together; we are going to Benjamin's residence," he said, gazing around the library. "I don't know what's going on with you, but I am going to find out," he whispered to himself.

A little while after, Phillip had made it to Mr. Benjamin's place, and it wasn't anything that he expected. There were metal air ducks around the entire house, and there were bushes lined up right underneath the ducks. There were motion sensors on the top of the house, with at least 3 on each side. There were also spotlights over each sensor. It almost looked like a fortress, with all the metal trimming. There weren't any door handles or windows. There were only slits about 10 feet off the ground like military bunkers. It was if he was prepared for anything. When Phillip made it to the front door, which was covered with metal plates. He was surprised that there was a doorbell. Since everything else was so phenomenal, the doorbell seemed as if it didn't belong. When he rang the

 Extended Version

doorbell, he didn't hear a sound. He wasn't sure whether it was because of how old the house was or because the bell didn't ring. Whatever it was, he heard the lock clicking as someone opened it from the inside. The door made a loud thump as it completely unlocked. It startled Phillip, so he jumped back, placing his hand on his gun. His eyes were beady, and he was ready for anything to happen at the moment. His pressure was already high from the string of events that had already taken the course, and he was determined not to let anything happen without any consequences. The day was growing old, and things were growing stronger by the moment.

The door had finally started to open up, and he heard a voice. "What do I have the pleasure of doing for you, Officer Phillip?" asked the voice.

Extended Version

# Replenishment

Phillip didn't respond right away; it took him a couple of seconds to get his heart to stop racing. His eyes had to gain focus on the situation that was at hand. He started looking in between the door and noticed an eye and a chain hooked to the door from the inside; it wasn't a small door chain-the chain was the size of a dog chain. He took his hand off his weapon. "Benjamin, is that you?"

"Yes, how may I help you, Officer?" he asked once again from behind the steel plated door.

"I need your help; my partner disappeared," Phillip said, trying to hold in all his emotions. He didn't want to scare Benjamin away, so he kept it simple and straight to the point.

"What do you mean your partner disappeared?"

"My partner has disappeared; as in, I think something has taken him," Phillip responded, hoping Benjamin wouldn't respond in a manner that would

                    Extended Version

cause him to have to leave. Suddenly, he could hear the door creeping back shut. "Boooff!" The door shut.

Phillip thought he had closed the door in his face, so he turned to walk away; his heart had almost completely stopped beating, for he knew that he wouldn't be able to take on this alien alone. For one thing, he didn't know too much about the creators besides what he read on the blog; and two, he still had no clue on how he would be able to get his partner back. Just when he had lost all hope, he heard the chain scrapping across the inside of the door, and the door suddenly started opening. "Come in," said Benjamin.

Phillip didn't waste any time; he scrambled toward the door. "Thank you," he muttered with excitement. "I really appreciate you taking the time to talk to me."

Extended Version

# Replenishment

Once he entered the door, it immediately closed behind him like a door on an old tomb. When he got inside, it didn't look anything like he had thought it would. There was a bench hooked to the perimeter of the wall through the house, and light seeped into the slits that were trimmed throughout the house where he could see one or anything coming. The walls directly in front of the slits were colored black. Also, mounted on each wall was a case containing a rifle and a machine gun. Phillip was annoyed by everything that had happened this morning, but he felt totally safe in Benjamin's house. Benjamin had transformed the house into a fortress of solitude, so all Phillip's fears had gone out the window. Benjamin led him to a room where the walls were white, and it actually looked like a house. There were a couch and a few chairs, a TV, a bookstand, and other home items.

 Extended Version

"You can have a seat over there on the couch. Would you like something to drink?" Benjamin asked; he stood by one of the chairs, waiting for an answer.

"Sure, a glass of water would be fine. I did have a long walk," Phillip replied. He looked around and noticed that one of the walls was covered in newspaper clips.

Benjamin had left the room to grab him something to drink; as soon as he walked out of the room, Phillip eased up and walked over to the wall and started reading some of the clips. *Man missing, with no leads on his whereabouts*, read one clip. *Woman missing no leads on her whereabouts. Kid found in the house alone* read another clip. *Man missing; he never made it home after work. No leads on his whereabouts* read another clip. The more he read, the more he noticed that all of the people were

Extended Version

disappearing on different days throughout the entire month, and after the last day of the month, it ended. All those people went missing, and no one had answers to where any of them were.

This was like an unknown chain of events that seemed to be happening in the same month. Phillip also noticed some older clips from 25 years prior. The weird thing was that it was in the exact same month, and there were people going missing every day of that month as well.  All in the month of May-31 people disappeared. As Phillip was getting more into reading each and every clip, Benjamin came back into the room. "Yeah, it's weird, huh? People, just disappearing without a trace. People who had a perfect life, people with kids. They have no regret for anyone they take.

"Their families see them for the last time, then they are gone. No one gets a chance to say goodbye, I love you, or to even get one last hug. I feel sorry

 Extended Version

for the families; I totally feel their pain. I lost my partner 25 years ago; I searched high and low but couldn't find a trace of him anywhere. I know you're here because you lost someone, and you know that no one will believe what you have to tell them, so you decided to find me-on a hunch that I can help you. I can't make any promises, but I will do what I can in opening your eyes to the truth," Benjamin said, handing Phillip a glass of water, then taking a seat. "Come take a seat, and I will tell you all that I know and, hopefully, something will help you in some shape, fashion, or form," Benjamin said, flagging him to have a seat.

Phillip stopped in his tracks. "How do you know I lost someone?" Phillip asked, grabbing the glass of water and walking over to the couch, taking a seat.

"Your partner, right?" Benjamin asked.

Extended Version

# Replenishment

"Ya your right, my mind is distorted right now?" Phillip replied, taking a sip of water, and then placing the glass down on a hardwood table beside him.

"OK, well, ask me whatever it is that you came to ask, and I will give you the best answer that I can."

"I lost my partner this morning, and I don't know what happened to him. One minute he was there, and the next, there was a bright light, and he was gone–like he had just vanished in the middle of nowhere. The weird thing was that his clothes were on the seat after he had disappeared. I know I'm not a crazy man. I had called for backup, and before anyone was able to make it on the scene, he was gone."

"Did you see a black ball before he disappeared?"

"Yes, I did."

 Extended Version

"If that's the case, then he's gone. I know because that's the same thing that happened to my partner. One minute he was there, and the next, he was gone, and his clothes were left behind."

Phillip didn't respond.

"I know it's a lot to take in."

"What do you mean he's gone? He can't just be gone. He was lying there, right beside me. There must be a logical reason for all of this," Phillip said.

"I know this isn't something that you want to hear, but he's gone, and he won't be coming back. I'm not sure where to, but he's no longer here on earth. Something happens when that black ball lights up. Their soul is somehow taken from their bodies, and after that, the body disintegrates. I don't know how it's possible, or even why it happens, but I've seen it with my own two eyes. When my partner was taken, I was at a far enough distance that I could see his soul had been ripped from his body; then his corpse looked as if it turned into gray ash and floated off. I had never seen anything like this before in my life until the night of May 10th, 1995. I will never forget that night as long as I live. They usually hunt during the night; the combination of the darkness and their invisible ability makes it difficult to escape. That's the whole reason why the ball is black–so it can merge with the night. They come every 25 years, treating our planet like a farming ground. They come and take who they want; then they leave our planet for another 25 years.

"I call them Whiffers because when they move, it sounds like the wind is blowing; then bad stuff happens. They came 25 years ago, and now they are back to start the process all over again. I don't know what they want with us, or why they come to our planet, but I do know that when they come, people



Extended Version

start disappearing all over the world," Benjamin said in a serious and stern voice.

"This is some bullshit. Is there a way we can stop them or even try to fight back?" Phillip asked, taking another sip of his water.

"I have been researching and coming up with different stuff that has been keeping them away from me. I came across some night vision and heat sensor shades that allow me to see them in their invisible state. I have also installed fan ducks around my entire house because they hunt by freezing the elements. So, it created a way that makes them think their ability isn't working-by activating the ducks which blow wind over the bushes. They are in control when things stop because they can hear you and only you. If nature is too loud, then it throws off their concentration, making it harder to hunt us.

# Replenishment

They are very mysterious creatures," Benjamin replied.

"How do you know all of this," Phillip asked.

"When you seek revenge, the little things that you didn't really pay attention to, show you the truth after you embrace reality. Showing up clearer than you could ever imagine," Benjamin said.

"I'm still confused," Phillip replied, leaning on the edge of his seat, hoping that Benjamin would give him a little something more to work with.

"Just have an open mind to reality, and things will be revealed when you are ready to accept them."

Phillip was confused.

"Phillip, things aren't always what they seem."

"An open mind to reality? I still don't understand."

"You will."

"OK, well, thanks for the information. Oh, and can you do me a favor?" Phillip asked, looking up at

 Extended Version

Benjamin. Benjamin looked at him, giving his full attention. "Can you not mention to anyone that I stopped by to see you?" he asked, hoping that Benjamin would honor his request. Benjamin nodded in agreement.

"No problem," Benjamin said, getting up from his chair, "Matter of a fact, you might want to exit out the back door. Make sure you're careful out there; it's just about nightfall, and they hunt at night," Benjamin said, walking over to his side table, opening up the drawer, taking out a flashlight. "Here, take this; you will need it," Benjamin said, opening up the back door so Phillip could exit the premises.

"Thanks for the hospitality. See you around," Phillip replied, walking off through a cut in the back of Benjamin's house.

# Chapter 12

Encounter

***Flashback***

May 10th, 1995.

"Hey, Jeff! Grab me a coffee, will ya?" Benjamin asked, browsing through the chip aisle.

"Already got it, Ben," Jeff replied, walking out of the gas station store. He paid for his snacks and headed out to the patrol car. He looked back toward the store to see if Benjamin was on his way out of the store. That's when he saw a pulsing, glowing light behind the building.

Curious, Jeff walked toward the light. As he reached the back of the store, he realized it was dark. He pulled out his flashlight and scanned the area. The light reflected off something invisible, about 10 feet tall.

           Extended Version

"What the ...?" Jeff muttered, keeping his light aimed at the object. As he walked closer, the light ricocheted off the object, hypnotizing him. Jeff, still entranced, reached out to touch it.

Meanwhile, Benjamin came out of the store, looking around for his partner. He noticed the flashlight moving behind the building and walked around to investigate. As he turned the corner, his flashlight landed on an unexpected sight, an alien-type creature, 7 feet tall, with orange skin, holding a black sphere. The creature turned and looked directly at Benjamin. The creature was right behind Jeff as he was reaching out to touch the object.

"Jeff, no!" Benjamin yelled, but it was too late.

The creature vanished, leaving the sphere floating mid-air. Jeff's hand touched the object, triggering a weapon system that powered up, and a laser beam type of weapon appeared. It aimed right

Extended Version 

at Jeff's hand, and Benjamin watched in horror as Jeff's hand was blown off. The black sphere that was floating sucked Jeff's soul right out of his body, leaving only his clothes behind, which descended to the ground slowly.

Benjamin took cover behind the dumpster, shocked and scared. "What's going on? Did I just see what I saw?" he repeated to himself, trying to process the events. He couldn't shake the vivid vision: Jeff was looking at him as he touched the object, the creature looming behind him–and then, suddenly, Jeff vanished. The vision just kept replaying over and over in his mind. At that point, it was too late.

Benjamin was alone, and he didn't know what to do next. He was so scared that he didn't know if he should move or not. The last thing Jimmy wanted was to suffer the same fate as Jeff, to vanish at the creature's whim. He sat behind the dumpster, his



Extended Version

mind racing. *What just happened? Did I really see an alien? Did Jeff ...?* He couldn't finish the thought. He peeked around the corner, his heart racing.

The oval object was still there, pulsing with an otherworldly energy. The black sphere was gone, and the creature was nowhere to be found. Before Benjamin knew it, the object vanished. That frightened him even more; his heart felt like it had dropped to his stomach. He knew he had to act.

Benjamin slowly backed away from the dumpster, trying not to make any noise. He needed to get back to the patrol car and call for backup. He eased away, watching every step he took, trying to be as quiet as possible. As he reached the car, he grabbed the radio and called for assistance. "Officer down! Requesting backup at the gas station on 5th and Main!"

# Replenishment

The dispatcher's voice was calm and reassuring. "Units are en route, Officer Benjamin. Can you give me a situation report?"

Benjamin took a deep breath, trying to process what he had just seen. "Jeff ... my partner ... he's gone. There's some kind of ... alien ... thing ... and a sphere ... it's got a weapon ..."

The dispatcher's voice remained calm and was quiet on the phone for a minute. "Officer Benjamin, what?"

Benjamin took another breath and responded, "Uhhh ... I don't really know ... my partner is gone ... I just ... need help."

The dispatcher's voice remained calm. "Help is on the way."

Benjamin looked around, feeling vulnerable. He felt like he was being watched. He knew he had to get out of there, but he was so nervous and shaky that he was scared to drive. Suddenly, he looked

 Extended Version

back, and, in his mind, he could see the creature staring back at him. He could not sit there any longer. He started the engine and sped away from the gas station, leaving the horror behind. But as he drove, he just got more and more paranoid. He felt like the creature was watching him. He could feel it was still out there, waiting for its next victim.

Questions kept echoing in Benjamin's mind. *What happened? What did I just see? Am I going crazy?* He couldn't believe what he had just happened. Jeff, his partner and friend, was gone. Taken by some kind of alien creature with its deadly sphere.

As Benjamin drove, he noticed a strange glow in the rearview mirror, but it turned out to be headlights from a car behind him. He was so paranoid that he floored it, weaving in and out of traffic. But no matter how fast he went, he still felt overwhelmed.

Suddenly, the radio crackled to life. "Officer Benjamin, this is dispatch. We've got multiple units en route to your location. Can you give us an update on your situation?"

Benjamin hesitated, unsure of what to say. How could he explain what he had seen? "I feel like I'm being followed by ... something. But I'm not sure ..."

The dispatcher's voice was skeptical. "Officer Benjamin, can you confirm that?"

Benjamin gritted his teeth. "I'm telling you, something strange is going on! Jeff's gone, and it's still out there!"

There was a pause on the other end of the line. "Officer Benjamin, we're going to try and get you to a safe location. Can you make your way to the nearest police station?"

 Extended Version

Benjamin nodded, even though the dispatcher couldn't see him. "Yeah, I'm on my way." As he drove, he just kept looking in his review mirror. He was trying to make sure the creature wasn't coming for him. Suddenly, an idea struck him. He remembered a narrow alleyway that ran behind the police station. If he could make it there, he could maybe have a little relief and perhaps peace of mind.

Benjamin took a sharp turn, heading toward the alleyway. He floored it, racing through the alleyway. In his mind, he could hear the pulsing sound of the object, remembering when he was behind the dumpster and its energy pulsing through the air. And then, just as suddenly as it had started, everything was silent.

Benjamin slowed down behind the station, shaken and alone as he pulled into the police station, his heart still racing. He knew he had to report what

Extended Version

he had seen, no matter how crazy it sounded. But as he stepped out of the car, he couldn't shake the feeling that he was being watched. The question popped into his head again and kept echoing repeatedly: *What happened to Jeff's soul?*

Benjamin walked into the police station, still trying to process what he had just seen. The desk sergeant looked up at him, concern etched on his face. "Benjamin, what's going on? You look like you've seen a ghost."

Benjamin took a deep breath, trying to gather his thoughts. "Jeff's gone," he said, his voice shaking. "He was taken by some kind of creature. It had some sort of black sphere. It killed him, then vanished."

The desk sergeant's expression changed from concern to skepticism. "Benjamin, slow down. Let's start from the beginning."

Benjamin nodded, trying to calm down. He told the sergeant everything: the gas station, the



Extended Version

glowing light, the oval object, the sphere, and Jeff's disappearance.

The sergeant listened; his expression unreadable.

When Benjamin finished, the sergeant nodded thoughtfully, even though he had a look of confusion on his face. All he could do was look at the blood that was all over Benjamin. But he didn't want to get Benjamin more on edge, so he said, "Okay, Benjamin. Let's take this one step at a time. Let's get you debriefed. But first, let's get you checked out by medical. You look like you're in shock."

Benjamin nodded, feeling a sense of relief wash over him. Maybe he was going crazy. Maybe it was all just a hallucination. But as he walked to the medical room, he couldn't shake the feeling that something was off. He didn't know if his story was believable. Benjamin dreaded being blamed for Jeff's

Extended Version

disappearance, while the true culprit, that monstrous creature, was still on the loose. The question kept echoing in his mind: *What happened to Jeff's soul?*

As Benjamin sat in the medical room, being poked and prodded by the doctor, his mind kept wandering back to Jeff. He remembered their partnership, their jokes, their late-night conversations. He was so lost with Jeff being gone. And then, he remembered something else. The last thing that Jeff did was touch the object, and his hand was blown off. Benjamin's eyes snapped open. He was too late; he was right there and could have saved his partner.

Benjamin's eyes locked onto the doctors. "I remember something," he said, his voice urgent. "There was some type of object, and it had a weapons system that triggered once Jeff touched it ..."

 Extended Version

The doctor's expression changed from curiosity to concern. "Benjamin, I think you're starting to remember things. That's good. But we need to take this slow. You've been through a traumatic experience."

Benjamin nodded, but his mind was racing. He just couldn't help but feel bad; his partner was right there, and he couldn't do anything to save him. Just then, the door burst open, and a tall, imposing figure strode in.

"Benjamin, I'm Agent Thompson from the FBI. I've been assigned to your case."

Benjamin's eyes narrowed. "What case?"

Agent Thompson's expression was grim. "The case of your partner's disappearance. And the ... unusual circumstances surrounding it."

Benjamin's gut tightened. He knew something was off. "What do you mean?"

Extended Version

# Replenishment

Agent Thompson hesitated. "Let's just say there's no evidence to support your story. So, as of right now, you're the only suspect."

Benjamin's eyes locked onto the agent's. "Suspect?"

Agent Thompson leaned in, his voice low. "So, you have the right to remain silent. Anything you say or do can be used against you in the court of law. You have the right to an attorney. If you can't afford one, then one will be appointed to you. Can you stand up and place your hands behind your back?"

Benjamin's mind reeled. He had suspected it, but to hear it confirmed just killed him inside. His partner was gone, and no one believed anything that he said. He stood to his feet and placed his hands behind his back. "Wait," he said, his voice urgent. "Never mind."



Extended Version

Agent Thompson's expression changed from aggressive to serious. "Yeah, it's best that you do not say anything."

Benjamin's eyes went from looking at the agent to the doctor. "He was my partner ... Please ..."

Agent Thompson nodded, a little hate in his eyes. "Let's go." He grabbed the cuffs and led Benjamin out of the room.

Extended Version

# Chapter 13

Court Day

The day had finally come for Benjamin to go to court, and he was nervous and angry at the same time. His partner was gone, and no one believed his story. It was just crazy to him that he was a cop and couldn't even get support from his fellow officers.

Benjamin was walked into the courtroom and placed in a seat beside his lawyer. Benjamin's eyes scanned the courtroom, his heart racing with every tick of the clock. The only face that brought him any joy was his ex-wife. She was the only one who had his back, no matter what. He was about to face justice for a crime he didn't commit.

The Judge entered the room, and everyone stood to their feet until they were told they could

        Extended Version

have a seat. The courthouse was full that day, but Benjamin didn't have one family member there for him. He was all alone, and now he didn't even have his partner to have his back.

The prosecutor called Benjamin to the bench.

Benjamin was sworn in and ready for questioning.

"So, Officer Benjamin," the prosecutor began, "you're saying that an ... alien took your partner's soul?"

Benjamin nodded, his voice firm. "Yes, that's exactly what I'm saying. It happened behind the gas station. I couldn't stop it."

"So, what did this so-called alien look like?" the prosecutor asked.

"Well, it was really dark, but it had these big eyes. It didn't have a mouth, and it was an orange type of color," replied Benjamin, very certain of what he was saying.

"Where was this alien creature at?" the prosecutor asked.

"Well, it was standing right behind Jeff when I came to the back of the store," Benjamin said.

"So, why were you behind the store in the first place?" the prosecutor asked.

"I was looking for Jeff; he wasn't by the car when I came out of the store," Benjamin replied.

"So, why were you in the store by yourself?" the prosecutor asked.

"I wasn't …," Benjamin replied, but before he could finish, he was cut off.

"Well, I thought you said you were looking for Jeff. So, if you were looking for him, how were you not alone?" the prosecutor asked.

Benjamin had a puzzled look on his face. "I was …"

 Extended Version

"But you just said that you weren't alone, and now you're saying that you were alone. Which is it?" the prosecutor asked.

Benjamin's lawyer stood up. "Objection, your honor."

"Sustained," replied the judge. "Let him answer."

Benjamin looked at his lawyer, confused. "Jeff and I were in the store together; then, he left the store after he got what he wanted."

"So, where did he go when he left the store?" the prosecutor asked.

"To the car, I suppose."

"So, was he at the car? Or was he behind the store? I'm just trying to get an understanding of where he was."

# Replenishment

"Look, I don't know. He could have been by the car; he could have been behind the store. All I know is, my partner was taken."

"Taken by who? There's no surveillance of anyone there with you and him." The prosecutor pointed a remote at a TV in the courtroom showing him and Jeff behind the store, but there wasn't anything else behind the store with them. Jeff was standing there, with his back turned from Benjamin, and there was no footage of a ship or an alien creature. When Jeff turned to look back, the picture got distorted, and when the picture came back into focus, Jeff was gone, and Benjamin was sitting on the ground behind the dumpster, looking like he was in shock.

Benjamin looked so confused. "But … There was something else there … I didn't kill my partner," Benjamin replied in a state of shock.

 Extended Version

# How Deep Does It Go

FBI Agent Thompson snorted. "Save it, Benjamin. We have evidence. Jeff's blood is all over your clothes."

"Order in my courtroom! Keep your comments to yourself, Agent Thompson," the judge said.

Benjamin's lawyer, Mr. Jenkins, stood up. "Your honor, my client is clearly distraught. I request a psychiatric evaluation before proceeding with the case."

The judge looked at Benjamin, who was frigid and had a frozen look on his face like he had seen a ghost. He looked back at Mr. Jenkins and nodded, and Benjamin was sentenced to a mental hospital for evaluation.

Days later, as he lay in his hospital bed, Benjamin couldn't shake the feeling that the creature was still out there while he was locked down in the facility. All he could do was think about

Extended Version

that night repeatedly. He just couldn't believe that his partner was gone.

Dr. Lee walked in and sat beside the bed. "So, you're saying that you saw some sort of creature?"

"Dr. Lee, I'm telling you; I saw it," Benjamin said, his voice urgent. "It was tall, with orange skin, and it was holding a black sphere. The black sphere was hovering when the alien disappeared; then the sphere shined so bright that I couldn't see, and when I could see again, it had taken Jeff's soul, leaving his clothes falling to the floor."

Dr. Lee nodded thoughtfully. "I believe you, Benjamin. But we need to find a way to prove it. There was no one else there or any footage to give your story credibility."

Benjamin's eyes narrowed. "I know it's out there. And I'll find it somehow, someway. It will strike again, and I hope I'm there to bring it down."



Extended Version

Benjamin just sat there; his eyes fixed on the wall as he relived the events of that fateful night. He couldn't shake how the creature looked back at him and how it just vanished like it was never there. The ship-typed object was something that he had never seen before. The pulsing sensation that it let off was something not of this world.

"What happened after the creature had taken Jeff's soul?" Dr. Lee asked.

"It was just gone; I was so shaken up about it that I sat behind the dumpster for a while, trying to make some sense of what I had just seen," Benjamin said.

Dr. Lee nodded thoughtfully. "And what did you do after that?"

Benjamin's eyes dropped. "I ran. I fled the scene and ended up back at the police station with Jeff's blood all over me."

Extended Version

# Replenishment

"OK, let's go back to that part. How did Jeff's blood get on you?" Dr. Lee asked.

"Oh, I forgot to tell you about that part. So, when I made it to the back of the store, I saw Jeff reaching out to touch this ship-shaped object. The object was pulsing, and Jeff was just about to touch it when I called him. He looked back, and I guess his hand touched it. When he did that, I guess it must have triggered some sort of weapons system because a laser beam blew off Jeff's hand, and his blood flew all over me," Benjamin replied.

Dr. Lee's expression was sympathetic. "I see. And do you know why you think this creature took Jeff's soul?"

Benjamin shook his head. "I don't know. All I know is, when I was able to see again, I saw a white ghost-like thing getting pulled from Jeff's body. It looked like Jeff was trying to fight it, but the sphere was too powerful, and it extracted his soul right

                    Extended Version

from his body. To me, it looked like a soul; I don't know what else it had taken from his body that could have been white. I know the creature is still out there somewhere."

"I know that must have been tough to witness," Dr. Lee replied.

"Yeah, it still is tough. Knowing that I couldn't save my partner," Benjamin said.

"I understand. How does that make you feel?" Dr. Lee asked.

"It makes me feel worthless. I was right there, and I couldn't do anything to save him," Benjamin explained.

"I know that must be tough on you," Dr. Lee said.

"You have no idea, doc," said Benjamin, but before he could finish what he was saying, the door

Extended Version 

suddenly burst open, and FBI Agent Thompson strode in.

"Benjamin, we need to talk," Agent Thompson said.

Benjamin's heart sank. "What now?" he asked, unsure of what was happening.

"Agent Thompson, what are you doing here?" Dr. Lee asked, her voice firm.

"I'm here to take Benjamin into custody," Thompson replied, his eyes cold. "We have new evidence that suggests he was involved in Jeff's disappearance."

Benjamin's eyes widened in horror. "No, that's not true! I would never hurt my partner."

"People say a lot of stuff, but we go by evidence," Thompson said with a look of spite.

Dr. Lee stood up. "Well, I won't let you take him. He's my patient now, and I won't let you harass

 Extended Version

him. So, you can leave; he's been court-ordered to be here."

Thompson sneered. "You can't protect him forever, Doc. He's going down for this."

"I didn't do anything. You're going after me when there's something else out there taking people," Benjamin replied.

"No, I'm going after the right one. His blood wouldn't have been on your shirt; he wouldn't be missing, and there would be some type of proof to validate your story. But there's nothing of that sort to help your case," Thompson said.

Dr. Lee didn't like Thompson interrogating her patient. "Agent Thompson, this isn't the place nor time for this. Justice has been served, so get a grip."

Agent Thompson wasn't too happy about that; he believed Benjamin was guilty, but he couldn't prove it at the moment. He just had to figure out

Extended Version

how to get Benjamin back in court so that he could get what was coming to him.

Benjamin looked at Agent Thompson. "I don't know why you hate me so much, but you're barking up the wrong tree. There have been plenty of cases of people disappearing, never to be found again. So, why are you so high-sprung about me when literally this is one of those cases? Open your eyes, Agent Thompson; there are beings out there who are a lot bigger, stronger, and smarter than us. Beings who are picking us off one by one."

Agent Thompson stood there for a minute; Benjamin seemed so sincere with his words, but he didn't believe in these beings that Benjamin kept rambling about. He went by tangible evidence, and Benjamin didn't have any. All he knew was that Jeff was gone, and Benjamin was the last person to see him.



Extended Version

"Today, you may be safe, but there is always a tomorrow," Thompson said with a look of determination on his face. "Justice will be served. Real justice. This isn't over ..." he said as he stormed out of the room.

Benjamin felt a sense of relief. He looked at Dr. Lee, then placed his hands together, resting his head on his fists.

"It's OK, Benjamin; today was a little much. We will pick up our session another day. Your nurse will be in shortly to give you some medicine to help you relax," Dr. Lee said.

As Dr. Lee left the room, Benjamin's mind raced. He knew he had to escape, to find the creature and clear his name. But how?

# Chapter 14

Obsession

Two years after Jeff's disappearance, Benjamin had finally come to terms with everything and learned to cooperate with the system, earning his release from the hospital. He was no longer in the force, of course, but he had a few friends who were still in the department if he needed anything, and he was certainly going to take advantage of his resources when the time was right.

Benjamin would be watched for a while, even though he was released. He was on probation for the time being. He had been confined in that hospital for so long; he had to get back in the midst of things. Even though he was out, he still felt like a little piece

 Extended Version

of him was missing. He missed Jeff more than anyone could understand.

Despite years of confinement, Benjamin counted himself lucky to still have a home to return to, thanks to his ex-wife's unwavering support. She had faithfully held onto the house and been his sole visitor during his hospital ordeal. Benjamin had saved her mother's life, and she was forever indebted to him. She didn't feel like it was an obligation; it was more of an appreciation thing.

No one could even understand why she had his back. Benjamin ended up giving her mother one of his kidneys after she had no luck finding one elsewhere. Benjamin's heroic act had created a debt she could never fully repay. So, while Benjamin remained in confinement, she kept his house clean and paid all his bills until he was able to return home.

Extended Version

# Replenishment

When Benjamin finally returned home, it was just like he had left it, and he was forever grateful. After he had taken a nice shower and got a proper sleep, he started his journey. He was determined to find out why the creatures existed and what they wanted.

Over the next few months, Benjamin sat at his cluttered desk, surrounded by newspaper articles, sketches, and notes. His eyes were fixed on a particular article that described a person who had gone missing and also people who had seen strange things, like the glowing sphere, the creatures themselves, or any unexplained weapons. "Whiffers," he muttered to himself, something that he had started calling them after he had read how, somehow, the creatures were able to freeze the elements and move around without making a sound. He had also noticed that they were referred to as Whiffers on different blogs throughout the internet.

 Extended Version

## How Deep Does It Go

Benjamin's partner, Jeff, had been taken by one of these creatures, and his soul was extracted by the retention ball, which was the name that Benjamin had given the sphere-type object. Benjamin had been driven to the brink of madness, ending him up in a mental hospital. But now, he was out and determined to learn everything he could about these so-called Whiffers.

Benjamin spent days poring over articles, looking for any mentions of the creatures or their technology. He had identified and sketched several of their weapons, including the Snatcher Mores, a claymore-like device that seemed to be used to capture human souls. He had also sketched out the Deaddimizer and Readdimizer-guns that seemed to be used by the Whiffers. Benjamin wasn't sure what they did, but people had mentioned the sighting of these weapons with a little visual description.

Extended Version

# Replenishment

Suddenly, Benjamin's phone rang, breaking his concentration. He had been going deep into his integration about the Whiffers. The call was from one of his old colleagues still in the force.

"Benjamin, I heard you're looking into some weird cases," his colleague said.

"Yeah, I am," Benjamin replied. "I'm looking for anything that might be linked to missing people or anything that's unexplainable."

"Well, I might have something for you," his colleague said. "We've had a few cases where clothing has been left behind, with no idea where the person's remains are, and witnesses have reported seeing strange, unexplainable things."

"That sounds like some interesting activity," Benjamin said, his excitement growing. "Can you send me the details?"

 Extended Version

"Sure, just keep this between me and you. I can get into big trouble if anyone finds out about me leaking information about any of these cases," his colleague said.

"Of course, I would never jeopardize your job," Benjamin replied.

"OK, well, I will get the information to you shortly," his colleague said.

As he waited for the information to arrive, Benjamin couldn't help but think about Jeff. He had lost his partner, his friend, to the Whiffers. But he was determined to learn more about them, to find a way to stop them. While he was researching the creatures, he was also adding certain upgrades to his house. He wanted to be prepared for anything. He was determined not to end up in Jeff's situation.

Benjamin spent countless hours looking over everything he thought would lead to anything

Extended Version

significant. He knew he had one chance to stop the creatures. He immersed himself in research, a familiar frenzy he hadn't experienced since his high school days. He had many sleepless nights; his appetite dwindled after a day or two, and he had to force himself to eat something. The whole situation brought a lot of stress, so much that you could see it in his waist size.

About a week had passed, and while Benjamin was involved with all his endeavors, the information he was waiting for arrived. He pored over it, his mind racing with possibilities. He was getting close; he could feel it. He just had to stay focused and see where it led him.

Suddenly, an idea struck Benjamin. He had run into some information online and wanted to see if he could get some information that might not have been on his radar. He had run across a guy who talked a lot about these creatures and was open to



Extended Version

meeting up with friends. He never gave a name or picture for privacy reasons. So, Benjamin grabbed his phone and dialed a number.

"Hello?" a voice answered.

"It's Benjamin. I need to talk to you about the Whiffers," Benjamin said.

There was a pause on the other end of the line. "What do you know?" the voice eventually asked.

"I know they're real," Benjamin replied. "And I know they're using these black spheres that I call 'retention ball' to extract human souls."

The voice on the other end of the line was silent for a moment. "I'm open; we can talk," it said finally.

Benjamin nodded, even though he knew the person couldn't see him. "I'm ready," he said.

Extended Version

# Replenishment

"OK. Well, we can meet you at the old warehouse on Omar Avenue at midnight," the voice said. "Come alone."

Benjamin nodded again, his heart racing with anticipation. He had been waiting for this moment for so long. Being able to talk to someone without being judged. He had been judged every step of the way, but he didn't feel any judgmental vibe while he had been on the phone. No one knew how it felt for someone to lose their partner and not receive any justice, only getting blamed for something that they didn't do.

At midnight, Benjamin made his way to the old warehouse, his senses on high alert. The pressure from him knowing that the creature was still out there somewhere had him on edge. He entered the dimly lit building, his eyes scanning the shadows. He wasn't sure of what he was walking into or even

 Extended Version

what to expect. This was a new chapter for him, and he hoped it would go smoothly.

Benjamin stood there for a while before a figure finally emerged from the darkness. Benjamin was a little caught off guard, frightened, and on edge, knowing these creatures were still out there.

"Benjamin," the figure said. "I see you're still obsessed with the Whiffers as I was at a certain point in my life."

Benjamin nodded. "I won't stop until I find out what happened to Jeff."

"Jeff," replied the figure, unfamiliar with who he was referring to.

"Yes, my partner. He was taken by the Whiffers," Benjamin said.

"Oh, I see," said the figure." The figure stepped closer. "I can help you," it said. "But you have to be careful. The Whiffers are not to be trifled with."

Extended Version                    237

# Replenishment

Benjamin's eyes narrowed. "Who are you?" he demanded. "And how can I trust you?"

The figure smiled. "I'm someone who knows more about the Whiffers than you do," it said. "And I'm willing to share that information with you. If you're open to it."

Benjamin hesitated for a moment, then nodded. "Okay," he said. "I'm listening."

The figure began to speak, telling Benjamin about the Whiffers' technology, motivations, and weaknesses. Benjamin listened intently, his mind racing with possibilities. Anything that he thought would benefit him was fair game. He wanted to figure out if there was any way for him to get his partner back. So, he continued to observe everything that the figure said. Benjamin had done a lot of research, but some of the stuff intrigued him.

As the night wore on, Benjamin realized that he had stumbled into something much bigger than

 Extended Version

he had ever imagined. The Whiffers were not just random creatures; they were part of a larger conspiracy that threatened the very fabric of human society. The creatures were taking souls, but what was the ultimate goal? The souls had to be used for something, and he was sure it wasn't for them to eat. There had to be a bigger and more severe reason. It almost seems like they target specific people. Benjamin wondered why the creature didn't take him as well. *Why Jeff?* he wondered.

Benjamin's determination grew. He would stop at nothing to expose the truth about the Whiffers, no matter the cost. For him to do that, he had to learn everything that he could about them, from their weaknesses to their weapons and whether they had been on planet Earth or were from another planet. He had to gather some valuable and tangible information about the creatures so he could have proof to present to people to show that he was right.

The figure revealed a lot of valuable information that Benjamin could add to his collage at home. Even though he felt that he was still so far away from uncovering the truth, this was certainly a step in the right direction. There was still so much more that he wanted to know about the creatures. He even wondered if it was possible to catch one of them. He didn't know if that would ever be a possibility, but it was something that he was surely willing to try. In order for that to happen, he had to know the functions of all the creatures' weapons, what the creatures were capable of, and how he would be able to catch something that he couldn't see.

Benjamin left the warehouse in search of equipment that he could use to defend himself from the creatures. All he knew was that it would be him

 Extended Version

or the creatures if it came down to it. He got online, and based on the information he had gathered, he wanted to find an item that would counter the creatures' abilities.

Benjamin had to dig deep and reach out to a few people to get certain weapons or items that weren't available to the general population: Tier One machine guns, infer-red goggles, Semtex explosives, and some sort of cage that was made from a material that one couldn't find at a typical hardware store. Everything that had happened had turned Benjamin's frustrations into an act of war against the creatures.

# Chapter 15

Caged

***Benjamin's Hideout***

Benjamin stood tall, his imposing figure transformed into a fortress of advanced military-grade body armor and tactical gear. From head to toe, every inch of his being was encased in a cutting-edge ensemble designed for maximum protection and lethality. His face was obscured by a black, ballistic helmet with a sleek, aerodynamic design. The helmet's visor, made of reinforced polycarbonate, provided optimal visibility while shielding his eyes from shrapnel and ballistic threats. A built-in communications system, complete with a throat microphone and earpiece, allowed for



Extended Version

seamless communication just in case he got in a sticky situation.

A customized Level IV hard-plate vest covered Benjamin's torso, providing unparalleled protection against armor-piercing rounds. The vest's ceramic plates were strategically positioned to safeguard vital organs, with adjustable shoulder pads ensuring a snug, comfortable fit. Over the vest, he wore a tactical jacket with multiple MOLLE (Modular Lightweight Load-carrying Equipment) attachments, hosting an array of pouches and tools. He wanted to be ready for anything, and that laser that he saw during Jeff's encounter wasn't to be taken lightly.

Benjamin's arms were protected by articulated; flexible forearm guards made from durable, high-impact polymer. These guards allowed for unimpeded movement while shielding his arms from fragmentation and slashing attacks. His hands

Extended Version

were clad in tactical gloves with reinforced palms and fingers, providing grip, protection, and tactile sensitivity.

His legs were encased in durable, ballistic pants with built-in knee pads crafted from the same advanced materials as his forearm guards. The pants' multiple pockets stored ammunition, medical supplies, and other essential gear. Benjamin's feet were shod in heavy, lugged tactical boots with ankle reinforcement.

A wide, sturdy belt cinched at Benjamin's waist, supporting his primary firearm–a customized, rail-mounted assault rifle. Additional pouches and holsters attached to the belt held secondary firearms, magazines, and specialized tools. A drop-leg holster secured his sidearm, readily accessible for the high-stress situation. There weren't too many people who were hunting aliens.

# How Deep Does It Go

A utility knife hung from Benjamin's neck. His tactical watch featured GPS, altimeter, and compass functions. A medical kit and emergency communications device were strategically positioned on his person. He knew what the creature was capable of, so he wouldn't be caught slipping.

Benjamin's attire was a mottled mix of matte-black, olive drab, and urban gray, expertly designed to blend into shadows and confuse his alien target. The overall effect was that of a ghostly, mechanized warrior, invisible at night. Benjamin figured he had to blend in as much as possible while facing the unknown.

As Benjamin checked his gear, his movements were fluid and deliberate, honed from years of training and combat experience. His eyes gleamed with a fierce determination burning beneath the armored visor.

With every piece of gear meticulously checked and double-checked, Benjamin exuded an aura of unyielding resolve. He was a finely tuned instrument of war, ready to face this creature head-on. "I'm coming for you, and I will find you," he whispered to the darkness.

Benjamin, an armored behemoth, moved forward with unstoppable force, driven by a fury of vengeance and anger. He then hunched over, listening intently to the police radio chatter.

Benjamin had been waiting for months, trying to get some sort of lead so he could track down this alien creature. He had shown up too many different calls with no luck. He had almost started to get discouraged that he couldn't track this creature down. Another call had come through, and he was



Extended Version

hoping that this would be the moment like he had hoped all the others would be.

Benjamin gathered some military-grade weapons with a nearly indestructible cage he hoped to trap the creature in. He was determined to have some vengeance for his partner. The more he thought about Jeff being gone, the more it ate him up inside daily.

"... Dispatch, we have a 10–96 at the old store on 5th and Main. Reports of bright flashes and sightings of an unknown entity ..."

Benjamin's eyes snapped to life. He did one more check and then sprang out the door. He jumped in his truck, which had a grand cage in its bed. He sped off into the night.


*** Abandoned Store ***


Extended Version

# Replenishment

Benjamin pulled up to the scene. He took the cage down and set it up by a side door, with the cage door open. He attached it to some sort of battery pack; then, he threw this huge sheet-type material over the cage and proceeded to enter the building.

Benjamin crept through the dimly lit store. He had his infrared goggles ready. Suddenly, a blinding flash illuminated the room.

"Jeff!" Benjamin shouted, charging forward. He proceeded to look around, hoping that his partner was there. The flash was eerily familiar–the same he'd witnessed the night Jeff vanished. It was a big hope, but he wished for the best.

Before Benjamin knew it, a figure materialized before his eyes–the alien creature. Benjamin opened fire, but the creature vanished.

"No!" Benjamin growled, fumbling for his goggles.

     Extended Version

# How Deep Does It Go

The goggles revealed the creature's heat signature fleeing through the store. Benjamin pursued, rifle hot and ready. Every time he thought he was getting close, he would see a flash, and when he made it around the corner, it would be a dead end.

"How in the world does it keep getting away!" Benjamin muttered in confusion.

No matter how many times this happened, Benjamin didn't let it deter him. He kept pushing forward. He had a plan, so he set up a trap to get the creature to move closer to the cage he had set up outside. He placed motion sensor mines at different locations to try to alter the creature's path and force it toward the side door.

After what seemed like an eternity, Benjamin's plan finally began to work. He could hear the mines going off and a screeching sound coming from the creature as it tried to maneuver throughout the store. He had finally started to make progress, and he was able to corner the creature. When he saw the creature, he took aim and fired his rifle, but he missed. The creature fled around the corner and into the cage. Benjamin advanced cautiously toward the cage, rifle at the ready, his goggles revealing the creature's thermal signature. With a swift motion, he slammed the indestructible cage shut.

Once the cage door was shut, Benjamin removed his goggles and pulled off the huge sheet he had over the cage. He looked at the empty cage, placed the goggles on, and looked to make sure the creature was still there, and he could see it.

"That is crazy. You can actually disappear. Wow!" Benjamin said in amazement. He had almost forgotten why he was there; he had to get some questions answered if possible. "You're going to tell me where Jeff is," Benjamin snarled.

 Extended Version

# How Deep Does It Go

The creature was in its invisible state, but Benjamin's goggles allowed him to see its shifting forms-a snarling wolf, a slithering snake, a flapping bird. The creature's attempts to escape were futile, as the cage had an invisible shield that stopped it from escaping.

"You can actually change form; no wonder you're so stealthy," Benjamin muttered. He proceeded to talk to the creature again. "Where is he?" he demanded.

The creature remained silent.

Benjamin went back to his truck, got on his radio, and told an approaching unit to stand down, that nothing was going on at the store-so he could have more time to interrogate the creature. He made his way back to the cage, brought a chair, and sat down by the cage with his goggles on, watching the creature and asking it questions until he was tired.

Extended Version 

# Replenishment

"I guess you can't understand me, huh?" Benjamin said.

Benjamin grew tired from hours of talking to the creature; before he knew it, a piercing hum filled the air. He looked around, trying to figure out what was going on. From afar, he noticed a pulsating entity floating closer toward him. The creature's ship materialized; a glowing band around it turned red like it was about to take fire; a hatch opened.

"No ... no, no, no," Benjamin muttered. He really wasn't expecting this to happen.

The ship's laser cannons swiveled, targeting Benjamin. He ducked and dodged the lasers. He dove behind walls as the laser cut through the walls like a knife through butter. A piece of the helmet that he had on was cut off at the top as he was crouched down behind the wall. After the wall was cut, it started to fall toward him. He hurried and leaped out of the way. The ship continued firing; every time he

 Extended Version

tried to peep and see if the ship was gone, it fired upon him again. He saw the ship powering up again to fire.

"Oh shit!" Benjamin shouted, grabbing his rifle; he jumped up and started running for his life. He sprinted away as the ship's cannons blasted at the store, trying to kill him. The ship was shooting so rapidly that he couldn't shoot back.

After Benjamin had escaped, the ship aimed at the cage and blasted the hinges off the door of the cage, and the creature escaped, fleeing into its ship and flying off into the night.

Benjamin was scared that the ship was still out there, so he sat in a corner, with his hand on his rifle, rocking back and forth, upset about the outcome.

*** Benjamin's Hideout ***

Extended Version

# Replenishment

Benjamin collapsed, defeated, the creature's cage empty beside him. He wasn't happy at all about the situation. "I had you," he whispered, staring at the cage. "I had you." He was really taken aback; he had done so much research and preparation for that day, and he couldn't make it work. He had a duty to avenge his partner. He was the only one who actually knew what happened, and no one could avenge Jeff but Benjamin. He just sat back and looked at his collage on the wall. He was prepared for everything but the ship. He had no idea that it would show up.

Days passed, and Benjamin's anger and despair deepened. He really beat up himself about it. He didn't know if he wanted to continue looking for more information about the creature. Clearly, there was so much more that he didn't know and

     Extended Version

had to learn in order to take a real stance against the creature.

"How could I have been so stupid?" Benjamin lamented, slamming his fist on the workbench. "I can't believe this." His eyes blazed. "Jeff's soul is still out there. I'll find another way."

Benjamin's determination reignited, but the weight of his failure hung heavy. He had a picture of Jeff on the wall; all he could do was look at it. He was fresh out of ideas but didn't know if he was ready to throw in the towel.

"I'll get you," Benjamin vowed to the creature. "Jeff, I'll get you back. Wherever you are." He looked at the sketch of the Whiffer. "You may have gotten away this time, but we will meet again, and when we do, I will be ready."

Benjamin stood motionless; his gaze lost in the vast expanse of the night sky. The stars twinkled like

Extended Version

# Replenishment

diamonds scattered across the velvet blackness, their gentle sparkle mesmerizing. His armor-clad body seemed to fade into the shadows, leaving only his contemplative spirit reaching for the cosmos. As he gazed upward, his mind wandered to the mysteries beyond Earth's atmosphere. The encounter with the alien creature had shattered his understanding of the universe. He had always considered humanity the pinnacle of existence, but now he knew better.

"How many of you are out there?" Benjamin whispered to the galaxy.

The silence was deafening. The stars offered no answers, but their silence sparked a torrent of questions within Benjamin. *Are there other civilizations like humanity struggling to comprehend their place in the universe? Or are they vastly different, their experiences and knowledge unfathomable to human minds?*

 Extended Version

# How Deep Does It Go

Benjamin's thoughts drifted to Jeff; his partner lost to the alien's soul-sucking sphere. *Is Jeff's soul still trapped, suspended in some ethereal realm? Or has it been transported to a realm beyond human comprehension?* The not-knowing gnawed at Benjamin's soul.

As Benjamin scanned the horizon, he spotted a constellation, Orion the Hunter. Its familiar pattern offered a fleeting sense of comfort, but his gaze soon drifted to the unknown regions beyond. *What lies hidden in the dark recesses of the galaxy? Other forms of life? Ancient civilizations? Forgotten technologies?* The possibilities swirled in Benjamin's mind like a maelstrom.

Benjamin recalled the creature's advanced technology, including the sphere, the Claymore device, and the egg-shaped ship. Humanity's most advanced achievements seemed primitive by comparison. He wondered if humanity was but a mere fledgling in the grand tapestry of existence. Certainly, when it came to the Whiffers.

# Chapter 16

In the dark

***Present day***

Back at Jimmy's house, he had just made it back home. The light from the falling sun grazed the treetops like a gentle breeze. He was so exhausted from running around throughout town that he paid no attention to the darkness that was taking its course.

He had been all over town, looking for his father, and he had called him several times throughout the day and still got no response. This was one of the most frustrating days of his life. Not only was he dealing with all the pains from the war, but his dad was also missing, and no one had a clue on his whereabouts. This was not like his father at all; Jimmy had a feeling in his gut that something

Extended Version

was really wrong pertaining to his father. He had exhausted all his leads, so he had no choice but to call his mother.

They were not getting along at the moment, but he didn't know what else to do. He plopped down on the sofa and placed his hands over his face, leaning back on the couch. A few seconds after, his wife walked in, and she could see that something heavy was on his mind. She walked over and took a seat beside him, placing her hand on his leg, rubbing it gently. "Baby, what's on your mind? Is everything OK?" she asked in a loving voice, caressing his leg continuously.

Jimmy inhaled deeply, looking at his wife; he was holding in a cry that was brewing deep down in his stomach, so he consumed his reply. A tear rolled around–under his left eyeball–testing the barrier of his eyelid. The fluids were slowly trickling over the edge, about to flood his pale cheeks like a Tsunami



Extended Version

in Thailand. She could see his emotional state peeking, so she quickly braced his head and pulled him close to her chest. "It's OK, honey. You don't have to talk about it. Just know I'm here for you," she said, feeling a little emotional herself for, she hated to see him like this. After feeling the warmness of her chest pressed up against his face, Jimmy uncontrollably busted out in tears.

"My father," Jimmy said, weeping heavily.

"What do you mean your father? Is he OK?" his wife asked, holding him tightly.

"I don't know," he muttered, grasping her arm. "I've been all over, searching for him, and no one seems to know where he is."

"He's a detective, so how does no one seem to know where he is?"

"I said the same thing; I don't know what to do. I was going to call my mother and see if she has heard anything," Jimmy said, leaning on her chest. "I know we aren't talking, but I have to at least try to reach out to see if she has heard anything."

"Well, I will let you handle your business, but if you find out anything, let me know. I will be taking a shower; if you need me, then I will come down and make dinner after I freshen up," his wife replied. She got up and headed upstairs to their bedroom.

"OK," Jimmy said as he reached over and grabbed his phone from the side table. He sat there a couple of seconds, looking at the screen, dreading to push dial on his mother's number. "Do I really want to do this?" he asked himself.

After a minute or so, he finally mustered the courage to push the dial, and he hoped for the best.

The phone rang several times; his mother finally picked up the phone. "Hello," she said.

Jimmy sighed. "Hello, Mom, I know we haven't talked in a while. Our last conversation wasn't too good, but this call isn't about me."

"Son, you left me at a really difficult time in my life to go into the army to almost get killed when I'm about to die myself. I thought I raised you with better morals than that. You completely turned your back on me when I needed you the most," Alice replied.

"Mom, I had a lot going on, and I just needed to get away. I have a life too, and I had to make decisions for myself. I might not always make the best decision, but I'm the one who has to live with them," Jimmy replied.

"You may not care, but I won't be here much longer. This cancer is killing me day by day. I don't have your father here, and I was really depending on you. If it weren't for your wife, I probably would not be here today. She was the only one making sure I got my treatments and taking care of me on my bad days. I would think you wanted to see me as much as you could before I wasn't here any longer, but I guess I was wrong. Your dad left me after our marriage of 10 years; then you turned around and did the same thing," said Alice.

"Mom, come on; you know why dad left you. He didn't leave you just for the fun of it. You have a gambling problem; all you did was gamble and smoke all day. You lost your whole life's savings in the casino. Even after that, dad tried to get you help, and you didn't want it.

He worked hard to provide a life for you and us, and you took that away from him. He even had money put aside for me, and you gambled that away. I would have been able to stay and go to school and get my doctor's. He had enough money put away for me that I could have paid my bills and paid for



Extended Version

school, but thanks to your problem; I had to figure something else out. You're holding something over my head that you created. I didn't call you to talk about that, though. I called to ask you if you had heard anything from dad today."

"I don't need this right now. No, I haven't heard from him, and I don't want to. I'm not going to sit here on the phone while you blame things on me. Bye," Alice responded, hanging up the phone in Jimmy's face.

"Mom," Jimmy replied, hearing the dial tone ring in his ear. He looked at the phone and shook his head as he hit the end button on his cellphone.

From the way things were going with his mother, he and she may never get back on a good foot; everything that was going on today with his dad only intensified. The conversation he just had with his mother only put a heavier damper on his frustration. Not only did he not get anything to help with his father's disappearance, but also, he and his mother only grew further apart.

He hated the fact that he and his mother had grown apart. He loved his mother, and he could remember the morning that he used to wake up, smelling those delicious maple syrup pancakes that she used to make every Sunday morning before they left out for the church.

After Jimmy's mother started her gambling addiction, all of that changed. She went from a mother that Jimmy loved with all his heart to a mother that made his eye twitch every time he looked her way. Her addiction cast a shadow over the love that her son once had for her, and she had no idea of the trauma it was causing in his heart. Not only was his love withering away, but also her marriage was on rocks, and her actions cost her everything.

 Extended Version

# How Deep Does It Go

While Jimmy was sitting back on the couch, dreading all that once was and still is, his wife walked in.

"Honey, how did it go?" Jessica asked, even though she was pretty sure that it didn't go well, for it never did. She gazed at his face and waited for him to reply.

Jimmy let out a deep sigh, looking up at the ceiling. "Not only did she go off on me and blame me for her problems, but she said that she hadn't heard anything from my father." His eyes started to water up slowly. "I'm so frustrated right now."

Jessica sat down beside Jimmy and hugged him. "Things will be OK, honey. I'm sure he's fine, and this is all just a big misunderstanding," she said.

Wiping the tears from his eyes, Jimmy said, "I sure hope so because this isn't like him."

# Replenishment

Right after Jimmy finished talking, Jessica jumped up. "I feel like I'm going to throw up." She put her hands over her mouth and ran out the room.

"Honey, are you OK?" Jimmy asked, getting up to follow her.

Suddenly, he heard a gust of wind outside of his window, but none of the trees in front of his window had budged whatsoever. He stood there for a minute and stared at the window as the greenery leaned on the window seal. Something had hit the aluminum trashcan, and it rolled across the lawn.

Jimmy reached over to his side table, reaching his hand in the drawer; he pulled out his sidearm, holding it snug by his side. He walks over to his front door; he peeped through the windows on both sides of the door. It seemed clear, so he proceeded out the front door. He was hoping that his father had knocked over his garbage can to get his attention. As

 Extended Version

he eased out through the front door, he whispered, "Dad, is that you?"

Jimmy walked out, closed the door, and looked around. He heard a gust of wind again coming from the side of his house. He raised his hand to see if he could feel any wind breezing through the air, but there was nothing. He was somewhat confused; he didn't know if his mind was playing tricks on him, or the wind was actually blowing, but he could feel the breeze. He proceeded to walk around to the side of his house to check out his residence. Again, he whispered, "Dad."

# Chapter 17

Enrollment

***Flashback***

Jimmy sat in his room, staring at the wall as he thought about the last 10 years of his life. He had spent most of his teenage years taking care of his mother, who was in and out of the hospital with various health issues. He never really had a chance to make friends or experience a normal childhood. He missed out on field trips, school rallies, and even the prom. He never had a normal childhood.

While at school one day, Jimmy was approached by a recruiter, and they had a long, interesting conversation about his future. Jimmy was intrigued to see that he had never opened his mind up to such things, and now he was eager to convince his parents that this was exactly what he

 Extended Version

needed. Jimmy thought long and hard that night about his options and whether this was the move he was ready to make.

Jimmy made up his mind and was set on moving forward with this ordeal. He decided to talk to his father first because he was a little more open about things. He was a lot less cutthroat than his mother. But Jimmy also didn't know the best way to approach the topic. He hoped it would be a little more natural, so he decided to wait for his dad to initiate a conversation. Then, he would just ease into the topic.

Jimmy was home relaxing the following day when his father dropped by for a visit. "Hey, Jimmy, how's it going?" his father, Phillip, asked as he entered the room.

"It's going, Dad," Jimmy replied with a shrug.

# Replenishment

"I know it's been tough, Son. But you've been a great help to your mom and me. We appreciate it," Phillip said, sitting down next to Jimmy.

"Yeah, I know. But sometimes, I feel like I've missed out on so much. I've never had any friends, never gone to parties or football games. I've just been stuck at home, taking care of Mom," Jimmy said, feeling a pang of sadness.

"I know, Son. But you've done what needed to be done," Phillip said, looking at Jimmy with concern.

"Well, I wanted to talk to you about something … I met a recruiter at school, and he told me about all the opportunities the army has to offer. I think it could be a good chance for me to start fresh and make a life for myself," Jimmy said, hoping his father would understand.

     Extended Version

# How Deep Does It Go

"So, you're thinking about joining the army? That's a big decision. So you're really thinking about doing this, huh?" Phillip asked.

"Yes, I think I need to start figuring out what I need so I can start building some type of career path," Jimmy replied.

"I understand, Son. And I'm proud of you for wanting to make a change. But you need to talk to your mom about it too," Phillip said, getting up and leaving the room.

Knowing he had to have the conversation with his mother, Jimmy sighed and got up. He found her in the living room, watching TV.

"Hey, Mom, can we talk?" Jimmy asked, sitting down next to her.

"What is it, Jimmy?" she asked, not looking at him.

# Replenishment

"I've been thinking about joining the army," Jimmy said, taking a deep breath.

Jimmy's mother's expression changed from calm to concerned. "Jimmy, I don't think that's a good idea. You're too young, and I'm not well enough to take care of myself if you leave."

Jimmy sighed. "Mom, you've been sick for years. I've been taking care of you since I was 10. When is a good time? You're never going to get better, and I need to start making a life for myself."

His mother looked away, avoiding eye contact. "I just don't think it's the right time, Jimmy. You need to focus on your education and get a good job. The army is not a safe place, and I can't bear the thought of losing you."

Jimmy shook his head. "Mom, I've thought this through. I want to serve my country and make a difference. And I need to do this for myself, not just for you or Dad."

 Extended Version

His mother turned to him, her face filled with anger and disappointment. "You're going to leave me? After everything I've been through. You're just going to abandon me like your father did?" she shouted.

Jimmy was taken aback by his mom's reaction. "Mom, that's not fair. Dad is here, and I'm not abandoning you. I just want to make a life for myself and do something good with my life," he tried to explain.

His mother's voice rose. "Your dad may be here physically, but he's checked out on me emotionally. If you leave, our relationship will be over. I can't lose you too. Your father hasn't been there for me, so if I lose you, I have nothing."

Jimmy felt a pang of guilt but stood firm. "Mom, I love you, but I can't stay here forever. I need to start living my own life. And if you can't support me, then maybe we do need some space."

Jimmy's mother's face fell, and she looked at him with tears in her eyes. "Fine, Jimmy. Do what you want. But don't come crying to me when you realize you've made a mistake."

Jimmy nodded, feeling a mix of emotions. He knew this conversation would never be easy, but he had made up his mind. He was going to join the army, no matter the consequences.

As he got up to leave, his mother called out to him. "Jimmy, don't do this. Please."

Jimmy turned around, looking at her with determination. "I'm sorry, Mom. But I must do this. For myself."

Jimmy's mother wasn't having any of it. "You're a selfish, ungrateful child. I can't believe you would even think of leaving me like this," she spat, getting up and storming out of the room.

 Extended Version

# How Deep Does It Go

Jimmy was left feeling hurt and confused. He had never seen his mother so angry before. He didn't understand why she couldn't understand his desire to make a life for himself.

Jimmy's fight with his mother had a lasting impact on him. She distanced herself from him, barely speaking to him, treating him like he was no longer her son. Jimmy was heartbroken, but he tried to move on, hoping his mother would someday come around and understand his decision. But as the days turned into weeks, and the weeks turned into months, Jimmy realized that his mother would never forgive him. She had made up her mind and was determined to treat him like a stranger.

Jimmy was still in denial about the outcome. He couldn't believe that his mother, the person he had spent his entire life taking care of, could treat him so cruelly. He hoped that one day, she would come to her senses and realize that he was still her son, no matter what choices he made. But until then, Jimmy was left to pick up the pieces of his broken heart and move on with his life.

The next day at school, Jimmy was still feeling a little down about all the commotion between him and his mother. He was trying to stay positive about the whole situation, so he scheduled a meeting with his recruiting officer.

Jimmy sat in the school cafeteria, drinking a soda, waiting for Jeff, his army recruiter, to arrive. He had just had a tough conversation with his parents about his decision to join the army and was eager to share the outcome with Jeff.

Jeff walked in, spotting Jimmy, and made his way over. "Hey, Jimmy! How's it going?"

Jimmy took a deep breath. "It's been a wild ride, Jeff. I talked to my parents about joining the army, and it didn't go so well with my mom."

 Extended Version

Jeff's expression turned sympathetic. "Sorry to hear that, Jimmy. What happened?"

Jimmy explained the conversation with his parents, detailing how his dad was understanding, but his mom fiercely opposed. "She's still not speaking to me, Jeff. But I've made up my mind. I'm joining the army, no matter what."

Jeff nodded firmly. "I'm proud of you, Jimmy. That takes a lot of courage. I'm sorry it's causing problems with your mom, but trust me, it'll be worth it in the end."

Jimmy smiled, feeling a sense of determination. "I know it will. So, what's the next step?"

Jeff pulled out a folder. "Well, first, you need to come by the recruitment office to get all your paperwork done and take your physical. Once that's complete, we'll start processing your enlistment."

Extended Version 279

# Replenishment

Jimmy nodded, taking the folder. "Got it. When do I need to be ready to leave?"

Jeff checked his calendar. "You'll need to report to basic training in six weeks. We'll give you all the details at the recruitment office."

Jimmy felt a surge of excitement mixed with nerves. "All right, I'm ready. Thanks, Jeff, for everything."

Jeff clapped Jimmy on the back. "You're welcome, Jimmy. Remember, it's not going to be easy, but it'll be worth it. You're making a great decision."

Jimmy left the cafeteria, feeling a sense of purpose and determination. He knew it wouldn't be easy, but he was ready to take on the challenge and start his new journey in the army.

Later, Jimmy made it home from school, and he wanted to talk to his mother one last time, hoping she would speak to him. So, after he had settled in,

 Extended Version

he walked to his mother's room and stood outside his mother's bedroom door, taking a deep breath before knocking softly. "Mom, can we talk?"

There was no response, but Jimmy knew his mom was awake. He had tried to talk to her several times since their argument about him joining the army, but she had refused to speak to him.

Jimmy opened the door and entered the room, approaching his mother's bed. "Mom, I need to tell you something. I'm leaving for the army in six weeks."

His mother turned her head away, her eyes fixed on the wall.

Jimmy continued, hoping she would listen. "I know you're upset, but I want you to know that I've made up my mind. I'm doing this for myself, and I hope you can understand and support me."

Extended Version 

# Replenishment

There was still no response, but Jimmy kept talking, hoping his words would somehow reach her.

"Mom, I know we haven't spoken in a while, but I want you to know that I love you and appreciate everything you've done for me. I'll be thinking of you while I'm away, and I hope we can reconcile when I get back."

Jimmy's mother remained silent; her expression unyielding.

Jimmy sighed and turned to leave, feeling a mix of sadness and frustration. He had tried to reach out to her, but it seemed like she was determined to cut him out of her life.

Jimmy felt a sense of resolve as he closed the door behind him. He knew he had done what he could to make amends, and now he had to focus on his future. He was joining the army and ready to



Extended Version

start his new journey, no matter how difficult it might be.

Jimmy walked away from the door, knowing he had tried his best to connect with his mother. He hoped that someday she would come around and they could rebuild their relationship, but for now, he had to move forward, no matter how hard it was.

The silence between them was deafening, but Jimmy held onto the hope that one day, they would find their way back to each other.

# Chapter 18

Tension

Jimmy was at school. Philip and Alice were standing in their living room, trying to have a conversation, but Alice had so much anger toward Philip that it progressed quickly, and their argument erupted into raised voices. The tension between them was palpable. Philip wasn't looking forward to having the conversation, but he knew it had to be done. They had been disconnected for a while now. Philip was hoping that they would be able to get back on the same page as they used to be.

"You're never here for me, Philip!" Alice shouted, her eyes brimming with tears. "You're always working, always putting your job first. I'm

 Extended Version

your wife, for crying out loud! I need you to be here for me, to support me."

Philip tried to placate her. "Alice, my job requires me to work awkward hours. I can't just leave in the middle of a case because you need me to be home."

Alice wasn't having it. "That's just it, Philip. You're always putting your job first. You're never here for me, for Jimmy. He's the one who's always there for me, not you. And it's not fair to him either. He's had to grow up too fast because you're never around."

Philip sighed, trying to reason with her. "Alice, I know it's tough, but I'm doing this for us, for our family. I want to provide for you and Jimmy, give you both a good life."

Alice just shook her head, her anger and frustration boiling over. "You call this a good life?

# Replenishment

Being married to a man who's never here? Having to rely on my son for support and comfort because my own husband is MIA? No, Philip, this isn't a good life. And I'm not sure I want to stay in this marriage anymore."

Philip's face fell, shock and fear written across his features. "Alice, don't say that. I'll try to do better, I promise. I'll make more time for you and Jimmy."

Alice let out a bitter laugh. "You've been saying that for years, Philip. And nothing ever changes. I need a husband who will be here for me, who will support me and our son. And if that's not you, then maybe we shouldn't be together anymore."

The fight ended with Alice storming out of the room, leaving Philip feeling defeated and unsure what to do next. He knew he had to make changes

     Extended Version

but didn't know where to start. And the thought of losing his wife and son was unbearable.

Philip tried to make more time for Alice and Jimmy as the days went by. He took days off work, attended school events, and helped with household chores. But the damage had already been done. Alice had made up her mind, and Philip knew he was fighting an uphill battle to save their marriage.

In the end, Philip realized that being a good husband and father wasn't just about providing financially but also being present and supportive. And it was a lesson he learned too late. Alice filed for divorce, and Philip was left to pick up the pieces of his shattered family.

Philip sat on the couch, his mind a thousand miles away. He was lost in thought.


*** Flashback ***

Extended Version

# Replenishment

Philip's thoughts drifted back to the day he first met Alice. That was one of the most unforgettable days of his life. She was a feisty one, always keeping him on his toes. But he was determined to win her over. Her beautiful smile hypnotized him; her voice was as soft as that of an angel. Philip was a delivery man then, and Alice worked as a receptionist at a doctor's office, so he often saw her. After a while, Philip had built up enough courage to ask her out.

Philip walked up and handed Alice a delivery. "How are you doing today, Alice? I was wondering, can I take you out for coffee sometime?" he asked, trying to sound suave.

"No, I don't think so," Alice replied, her eyes sparkling with amusement. "I'm not interested in dating right now."

"OK, I can respect that. Well, I hope you enjoy the rest of your day. I'll see you next time. See you

          Extended Version

later, Alice," Philip replied, smiling as he walked out of the office building.

Philip didn't give up. He kept asking her out, sending her flowers, and showing up at her work with coffee.

"Hello, Alice. How are you today," Philip asked.

"Hey, Philip. I'm doing fine. I see that you are a pretty persistent man; thank you for all the kind gestures. But you have to stop sending all these flowers to my job and bringing me coffee. I appreciate it, but ..." Alice said, trying to hold in her smile.

"I understand, and I was planning on stopping ..." Philip replied.

"Oh, OK," Alice said, not ready for the response that she had just gotten.

"As soon as you let me take you on a date," Philip replied.

Extended Version

# Replenishment

Alice smiled big; she was unable to hold it in any longer. "OK, Philip. I guess," she responded.

"Wait! Huh?" Philip said, happy about her response, "OK, cool. OK, a date it is," Philip replied.

Finally, after weeks of persistence, she agreed to go out with him. Their first date was a disaster. Alice was late, and Philip was nervous. But they laughed about it, and he took her out for ice cream to make up for it.

As the weeks turned into months, Philip and Alice grew closer. They went on long walks, had deep conversations, and explored the city together. Philip knew he wanted to spend the rest of his life with her. They were out on the town one night when he asked, his heart racing with excitement, "Alice, will you be my girlfriend?"

Alice smiled, and Philip's heart skipped a beat. "Yes, I'll be your girlfriend," she replied.

                    Extended Version

## How Deep Does It Go

A year later, Philip proposed to Alice, and she said yes. They got married, bought a house, and spent their days traveling and enjoying life. When Alice got pregnant with Jimmy, they were over the moon with excitement. They wanted a baby together so badly, and their dream was finally coming true.

Jimmy's birth was one of the happiest moments of their lives. Philip remembered holding his tiny son in his arms, feeling a sense of pride and joy he had never felt before. He wanted to be the best father that he could be. Philip didn't have much of a relationship with his father, so it weighed on his heart for most of his life. His mother was the only parent that he knew.

**** ****

As Philip sat on the couch, lost in thought, his mind snapped back to reality. All the thoughts that he had back when they had first met were just dwindling as if they had never existed. He thought about the argument he and Alice had just had and the pain in her eyes when she said she wanted a divorce. It really killed him inside, but he knew that his decisions over the years killed the light that once lit up Alice's heart.

Philip felt a wave of sadness and anger wash over him. He was angry at himself for letting their lives turn out this way. He had become so consumed with his job as a police officer that he had neglected his family and didn't know how to fix it. That was the most terrifying thing to him now. Knowing what was happening but unable to repair it, Philip continued beating himself up about it.

"Why did I let this happen?" Philip thought to himself, his eyes welling up with tears. "I was so focused on providing for my family that I forgot to be present for them."

 Extended Version

# How Deep Does It Go

Philip's mind raced with thoughts of what he could have done differently. He thought about all the missed birthdays, the late nights at work, and the times he had prioritized his job over his family.

As the tears streamed down his face, Philip knew he had to make a change. He couldn't lose Alice and Jimmy. He had to fight for his family, for the life they once had. It would all have been for nothing if he had lost everything so dear to his heart.

With a newfound determination, Philip got up from the couch, ready to face whatever challenges lay ahead. He knew it wouldn't be easy, but he was willing to do whatever it took to save his marriage and be the husband and father his family deserved.

Jimmy walked through the front door, went upstairs, and dropped his backpack on the floor. After showering, he headed to the table to have dinner. He could smell it from his room, and he was so ready to enjoy a home-cooked meal; he was growing tired of the cafeteria food. When he walked in, he could feel the tension in the air, like a thick fog that clung to every surface. Dinner had just gotten done, and his parents went and sat down at the dining table, their faces stern and unyielding. He sat at the table.

"Hey, kiddo," Philip said gruffly, not looking up from his phone.

Jimmy hesitated, unsure of how to navigate this awkward situation. "Hey, Dad."

Alice remained silent, her eyes fixed on some point beyond Jimmy's shoulder. He tried to catch her eye, but she refused to acknowledge him.

Jimmy shifted uncomfortably, feeling like an outsider in his own home. "So, how was your day, Dad?" he asked, trying to break the silence.

Philip shrugged, his expression unchanging. "It was fine."

 Extended Version

# How Deep Does It Go

The conversation was stilted, like a forced march through a minefield. Jimmy tried to think of something else to say, but his mind was blank.

Alice finally spoke up, her voice cold and detached. "Jimmy, can you eat your food and then go to your room? I need to talk to your father."

Jimmy nodded, feeling a pang of disappointment. He had been hoping for some interaction, some connection with his parents. But it seemed like they were too caught up in their own problems to even notice him. He just sat there and ate his food while silence pierced right through the room.

Jimmy finished his food, rose from the table, and placed his plate in the sink. As he walked out of the room, he glanced back at his parents before trudging upstairs. Jimmy felt a sense of despair wash over him. His family was falling apart, and he didn't know how to fix it. He felt lost and alone, like a small boat adrift in a stormy sea.

Jimmy flopped down on his bed, staring up at the ceiling. He wondered why his parents couldn't just talk to each other and why they had to be so angry and bitter all the time. He sighed, feeling a deep sadness settle in his chest. He just wanted his family to be happy again, to be like they used to be before his father became a police officer and his mother started being sick all the time. He just sat there, wondering why and what if things were different. *What if I didn't want to join the army? What if my dad never became a police officer? Would things be different?* Jimmy wondered. But as he lay there, he knew that was a pipe dream.

Jimmy's parents were too far gone, too caught up in their own problems to even notice him. He was left to navigate this treacherous landscape alone, searching for a connection that seemed farther away



Extended Version

than ever. He didn't understand how one decision could have such a big effect on one's entire life. He had to be smart about things he decided to do in life because he didn't want to get to the point where his parents were. He wanted better for his future.

# Chapter 19

The Divorce

Phillip and Alice's arguments had become a regular occurrence. They would fight about everything, but the underlying issue was always the same: Phillip's job. Phillip was so into trying to become a detective that he didn't realize the effect his job had on his family. Phillip and Alice were at the table, having breakfast, and the only thing that wasn't dim was the withering smoke from Alice's coffee.

For a moment, Alice was quiet, but she couldn't contain herself anymore. "Would be nice if you weren't always working," she said, her voice shaking with frustration. "You're always putting your job first. What about me? What about Jimmy?"

          Extended Version

# How Deep Does It Go

Phillip was a little annoyed about everything. "I'm doing this for us," he replied, his voice firm. "I'm doing this to provide a better life for our family. Well, at least that's what I thought I was doing."

Alice was tired of Philip's excuses. She was tired of being alone. It had been so many lonely days ever since Phillip came home excited to tell her about the opportunity that he had become interested in.

"But at what cost?" Alice asked, tears streaming down her face. "You're never here for us. You're always working, always chasing your dream. But what about our dreams? What about our family?"

Phillip sighed, running his hands through his hair. "I know I've been working a lot, but it's not forever. I'll make it up to you, I promise."

Extended Version

# Replenishment

Alice shook her head. "Promises, promises. That's all I ever get from you. You're always promising to change, but nothing ever does."

Phillip's face reddened. "That's not fair. I'm trying my best." He was lost and really didn't know how to respond to his wife. This was really a difficult situation, and he was somewhat getting discouraged about the whole thing. Alice was so furious that she was constantly on the attack. She continued to lash out.

"Try harder," Alice said, her voice cold. "Try harder to be a husband and a father. Try harder to be present in our lives." Even though she wasn't talking to Jimmy, she also found herself defending him.

Alice just stopped and got quiet. She just got to the point where she was trying to calm down. She felt like Philip wasn't listening, so she jumped back



Extended Version

into the shell she had been in for years. She started sipping her coffee, which was getting cold.

Suddenly, Phillip's phone rang, breaking the silence. He hesitated for a moment before answering. He knew it was his job, and it would probably get Alice going again. It rang again, and no matter how much he didn't want to answer, he didn't have a choice.

"Detective James," Philip answered, his voice professional.

Alice rolled her eyes. "Of course, it's your job. Your job always comes first."

Phillip ignored her as he shook his head, listening intently to the person on the other end of the line. After the conversation was over, he looked up at Alice. He dreaded telling her, but he just ripped the bandage off. "I have to go," he said, hanging up the phone. "There's been a break in the case."

Extended Version

# Replenishment

Alice threw up her hands. "Of course. You have to go. There has also been a break in your marriage, but your job always comes first again."

Phillip sighed. "Alice, please understand ..." he said, even though he knew she wouldn't.

Alice was drained, worn down by the relentless expectation to empathize. She had reached her limit. She was tired of being alone. She was tired of having a piece of her husband. She never dreamed they would get to this place in their marriage in a million years, but that dream had become dormant.

A few months later, Jimmy came home from the army on leave. He had joined the army to get away from the tension in the house. He loved his parents, but he couldn't take the fighting anymore. He was excited to see his parents, but his excitement was short-lived. Once he found out about his parents

 Extended Version

splitting up, he wasn't surprised. It was almost destined to happen.

Jimmy could feel all the bad energy in the atmosphere when he walked into the house. "Dad, what happened?" he asked, sitting down with his father.

Phillip hesitated. "Your mother and I ... we just grew apart, I guess. We wanted different things."

Jimmy nodded. "I felt like this was going to happen. You were always working, and Mom felt alone. She felt like you were choosing your job over her."

Phillip looked away, unable to meet his son's gaze. "I didn't mean to make her feel that way. I just got caught up in wanting better."

Alice was beyond consolation. She was tired of Phillip's apologies, weary of his excuses. She looked at Phillip and Jimmy while they were talking and

Extended Version

couldn't help but say something. "I don't want to talk to you, Phillip," she said, her voice cold. "I don't want to deal with you or Jimmy anymore."

Phillip tried to reason with Alice, but she was unmoved. He could see the frustration on her face; he really wanted to make things right. It was just becoming stressful because he didn't know how to. "Alice, please," he said. "I'll do anything to make things right."

But Alice just shook her head. "It's too late, Phillip. I'm done. I just want to get this divorce over and done with."

While Jimmy was in Germany, he tried to reach out to his mother, but she wouldn't even acknowledge him. So, he decided to try again while he was right there with her. "Mom, please talk to me," he said, feeling abandoned, but Alice just turned away, unable to find any words to say.



Extended Version

Phillip tried to make things right, but Alice was too hurt. She felt abandoned by both her husband and her son. Neither of them was able to get through.

As the days turned into weeks, Phillip realized that he had lost his wife. Someone that he loved most in the world besides Jimmy. Jimmy wasn't as hard as Alice, so Philip believed he could rebuild his and Jimmy's bond. He tried his best to make amends with her, but it was too late. The damage was done, and all he could think about was the divorce and how it seemed to be imminent. All he could do was hope that someday, somehow, he could make things right.

Phillip was heartbroken. He had never felt such a deep loss before. He had lost the love and connection with his family; his wife had withdrawn, and his son's admiration had waned. He didn't know how to move on.

Jimmy, on the other hand, was trying to come to terms with his mother's rejection. He had never seen her like this before. She was always warm and loving, but now she was cold and distant. Jimmy was conflicted about how to feel about his father. So, he sat down and decided to have a talk to try to get a little understanding of their relationship.

"Dad, what did I do wrong?" Jimmy asked, feeling tears prick at the corners of his eyes.

Phillip shook his head. "It's not you, Son. It's me. I'm the one who messed up. I put my job first, and I lost sight of what's truly important, and that's you and your mother."

Jimmy nodded, trying to understand. "But why won't Mom talk to me? I would think that she wants me to do better and achieve goals in life."

Phillip sighed. "I don't know, Son. I think she feels abandoned by both of us. She feels like we left her behind, which I understand. It's just hard to see her like this, knowing that it's my fault."

Jimmy felt a pang of guilt. He never meant to hurt his mother. He didn't purposely try to hurt her, but he had to make some changes in his life. Unfortunately, he ended up hurting her during the process.

As the days turned into weeks, Phillip and Jimmy tried to move on. They started attending counseling sessions, trying to work through their issues. The process was going quite well; they were starting to rebuild their relationship, whereas Alice was still distant. She wouldn't talk to Jimmy, and she barely spoke to Phillip.

Phillip knew he had to do something drastic to win his wife back. He didn't want to quit his pursuit to become a detective, so he didn't know what else he could possibly do.

Jimmy saw the change in his father and started to open up to him again. He didn't know if it would save his marriage, but it was something to help them move forward, no matter the outcome. After their counseling session was over, he started to feel a sign of relief.

"Thanks, Dad," Jimmy said, smiling. "I know you're trying."

Phillip smiled back. "I'll do anything to make things right, Son."

Alice was still a long way off and probably wouldn't change her mind about the divorce. It had just become so much and only got worse over time.

Months went by, and Phillip continued to go to counseling. He worked and went to counseling to find some peace with everything that was going on.



Extended Version

Even though Jimmy had left and returned to Germany, Phillip still sought guidance and support.

Jimmy checked in with his father more often than usual. He knew it would be tough for his father to lose the love of his life. He sent his dad letters so that he could continue to make sure he was OK and continuing his self-progression. It was tough on Jimmy also, as he could not talk to his mother like they used to.

Alice remained distant. She occasionally attended family gatherings but wouldn't engage with Phillip. She was polite but cold, and no matter how many letters Jimmy sent her, she would just ignore them and toss them in a pile that she started accumulating in the corner. It was starting to look like an abandoned post office.

Phillip didn't give up. He kept showing up and kept trying to make amends. He realized that

Extended Version

regaining Alice's trust would require patience and perseverance; earning a second chance would take time.

A few months later, Philip was served his divorce papers, and his heart dropped. He knew at the moment that he would never get Alice back, and his family would be forever broken. Seeing that letter put him in a slump for the rest of the day. He could not focus on work, so he was told to take a little time off to get himself together.

Philip tried one more time to reach out to Alice, but he didn't get a response. So, he had to just deal with the consequences of his actions, no matter how harsh and unfair they were.

After the divorce was all said and done, Philip ended up writing to Jimmy and explaining everything.

When Jimmy received the letter from his father, he opened it and started reading. His mood



Extended Version

immediately changed. He was really hoping that the outcome was different. But his mother was stern with her decision and unwilling to fix anything with him or his father.

# Chapter 20

Taken

***Present day***

Back in the house, Jimmy's wife had just finished throwing up. She grabbed a hand towel off the counter of the sink and wiped her mouth. She sighed deeply, holding her stomach tight as she laid back her head. "Oh my God, what's wrong with me? I haven't been able to hold down all day. I hope I'm not getting sick," she said and then slowly climbed from the floor.

She cut on the water and washed out her mouth, then dried it with the hand towel once more. Feeling a little better, she walked out of the bathroom to the living room, looking for her husband. "Jimmy where are you?" she called out, looking around dramatically. She looked in the



Extended Version

kitchen, and then she walked over to the stairway. "Baby!" she yelled.

Outside, Jimmy was on the side of the house, picking up the garbage can; he heard his wife calling for him. He placed his gun on the ground and placed the garbage can by the house, and when he turned around, there was a creature standing there, gazing right into his eyes.

The creature had a reddish-orange skin complexion. Its eyes looked like a three-edged star. It had no nose, and its mouth was like ours–but only a half of a mouth. It had a couple of bands on its wrist and some sort of tactical weapon holders with some sort of weapons. Its body resembled that of a human. It startled Jimmy, causing him to jump back, tripping over the garbage can; he fell to the ground.

Jimmy looked up at the creature, noticing that it had a black ball in its hand, which started spinning to spin slowly. The creature started to lower its hand, and the black ball stayed there, floating in the air. Jimmy looked down and noticed his gun not too far away.

He jumped over and grabbed his gun, and when he whipped around, pointing his gun, the creature had vanished, but the sphere was still there, spinning tremendously fast. It didn't look good, so he aimed and shot the sphere. The impact of the bullet caused it to spin out of control, smoking and crashing through his kitchen window. Jimmy had to jump out of the way because the ball was aiming right for him. While he was trying to get up from the ground, his wife ran over to him so she could help him up. "Baby, are you OK? What's going on?" she asked.

"I don't know, honey. I think I saw an alien," Jimmy said. He got to his feet; he could hear an



Extended Version

animal growling, and from the sound that the animal was making, it was defiantly big.

He slowly turned around, and there was a 600-pound grizzly bear standing a few feet away from them, with teeth like a barracuda, drooling like a Saint Bernard, staring at them like a midnight snack. The beast growled as slob trickled down its teeth. Its eyes grew dark, and its claws barreled deep into the rough gravel. Jimmy grabbed his wife and pulled her close to him slowly with his right arm as he held his gun in his left hand, ready to shoot right into the beast's chest.

Jessica eased behind her husband, and they tried to step backward away from the bear. For every step they took backward, the bear snarled and dug its foot into the ground, ready to charge at any minute. Their hearts thumped in their chests like a

Extended Version

concert slab that was dropped from a two-story building.

Their hearts beat harder and harder every time the bear dug its paws into the ground. While backing up, Jimmy stumbled, and before he could brace himself, the bear charged, and he fired his gun, hitting the bear in the shoulder. Green blood splattered on the ground, gushing from the bear's shoulder. The bear clawed at Jimmy, and his wife jumped in the way; the bear's claws scratched her arm, and its blood oozed in her cuts from the claw marks. Jimmy pushed her out of the way, and before she could turn around, the bear transformed back into the alien creature that it once was.

The alien pressed a button on its Flexure Truss, causing a portal to appear for a second so it could escape through the chestnut oak fence; it jumped back into the portal and flicked a small rubber ball the size of a dime that stuck right on the

 Extended Version

side of the house directly above Jimmy's head.  As soon as the portal closed, Jimmy stood to his feet to run toward his wife, and the rubber ball was trigged; a beam of light extracted Jimmy's soul right from his body. His soul didn't extract like the rest; the metal plate in his head altered its effect. Instead of him forgetting everything that had happened (family, name, wife, etc.), he could remember everything that ever happened to him throughout his entire life. Once his soul had left his body, what was left was a bright green instead of a dull white. His corpse disintegrated as his clothes drifted to the floor. His wife screamed as her husband vanished right in front of her face. She ran over and grabbed his clothes from the ground. "Noooooo! Jimmy … Someone help …" she screamed as tears flooded her face.

# Replenishment

About a hundred yards away, Phillip heard her screams for help and sprinted toward the house.

Jessica had Jimmy's clothes in her hands, crying, as Phillip ran in the direction of her cries.

"Jessica, is that you?" Phillip yelled as he approached the house.

"Phillip, is that you?" Jessica asked, confused, for she thought something had happened to him.

"Yes, where are you?" Phillip asked, looking around in the front of the house.

"I'm over here; they took him, Phillip. They took your son," Jessica yelled in a state of shock.

Phillip rushed to her. "Who took him?" he asked, holding her hands. "What happened?"

"I don't know. There was a bear; then I saw something that looked like an alien, and then he disappeared," she said in disarray, trembling in fear.

"What? Slow down, honey? What do you mean there was a bear?"

 Extended Version

## How Deep Does It Go

Jessica was shaking, trying to get herself together. Phillip wiped the tears from her face. "Jimmy and I were in the house, and I wasn't feeling well, so I went to the bathroom, and when I came out, Jimmy was gone. So, I started looking for him. I checked the living room, then the kitchen; suddenly, something came crashing through the window, so I ran outside, and I saw him on the ground. I rushed over to help him up, and as soon as he got up, we turned around, and a giant bear was standing there, growling at us. Then the bear scratched me, and Jimmy pushed me away, and when I turned and looked up, I saw an alien; then it vanished into a hole. Then Jimmy stood up, and there was a bright light; then Jimmy was gone, and his clothes were falling to the ground. I know this sounds crazy, but that's what happened, I promise. I just want Jimmy back," Jessica said, totally terrified.

Extended Version

# Replenishment

Phillip just stood there in shock for a second. "Oh no!" he replied.

"Oh no what?" Jessica asked, wanting to know what was going on.

Phillip dropped to his knees. "Not my son," he said; a tear ran down his cheek as he balled up his fist in denial. "No, not my son."

Jessica was totally lost, so all she could do was place her hands on his shoulders and cry as she continued to ask him over and over, "Oh no what? What's going on? What is it that you're not telling me?"

Unknown to them, Jimmy's soul was trapped inside the little rubber ball that was stuck to the wall behind them. He was calling out, "I'm right here, honey! Why are you guys crying?" he yelled, not knowing that his soul was crying out, and there was no ear that could hear the cries of a soul. Jimmy could see them as bright as day in this fish-eye-

 Extended Version

shaped ball, which he was trapped in. He couldn't move nor get their attention, no matter how much he screamed out for help. His soul circled around in this black ball, like a misty white smoke, which glowed every time he yelled out. "Help … I'm right here," he cried out as the mist formed a fist as he pounded on the interior of the ball. His voice sounded loud to him, but to bystanders, it wasn't even a whistle in the wind. After endless attempts to get their attention, he started to give up, for his hope started withering away.

Since Phillip wouldn't enlighten Jessica to what was going on, she started crying out in agony. Phillip was trying to hold it together for the both of them, so he put his arms around Jessica and said, "Honey, let's go inside, and I will try to explain to you what's going on, and I will tell you everything that I know up to this moment." They both rose to

Extended Version

# Replenishment

their feet and walked toward the front door. When they bent over the corner, a hand reached over and grabbed the small black ball off the wall that Jimmy was concealed in. The hand disappeared back through the small portal from which it had come, taking Jimmy with it.

Meanwhile, Jessica and her father-in-law were sitting in the living room, talking about and trying to figure out what was going on with Jimmy's disappearance. Jessica was sitting on the couch, trying to hold it together so she could hear what Phillip had to say about this terrifying event.

Phillip sighed deeply. "This morning, me and my partner were out, checking out a crime scene, and before we knew it, we had stepped into a war zone. Somehow, Wayne got shot, and I ended up dragging him back to the car drenched in blood. Suddenly, a black ball was hovering a few yards from the car; suddenly, it vanished, and I was so

 Extended Version

scared that I shot my gun, and I hit what seemed to be an alien," Phillip said. Jessica had a look of confusion on her face.

"What do you mean an alien?" Jessica asked, not sure if that's what he really meant.

"Yes, an alien. I was pretty sure that's what it was because nothing was there. Then I fired my gun, and I hit it. When I hit it, the creature appeared. Then it vanished again a couple of seconds later, and then ran off into the bushes. Before I knew it, the ball that was hovering appeared right in the back seat hovering over my partner; then there was a bright light, and when I looked up, he was gone. All that was left was his clothes. Wayne was gone, and I couldn't find him anywhere. Since I wouldn't have been able to explain what had happened, I left the scene because no one would have believed me. I'm pretty sure it's still hard for you to believe, even though you just witnessed the same thing with jimmy," Phillip said, still in a state of shock as he looked at Jessica.

"I can't believe this is really happening. I do hope that I'm dreaming, and I wake up any moment now. I just want my husband back, and I wish none of this ever happened," Jessica cried out in despair. Phillip held her and consoled her as he tried to figure out how to get his partner and his son back. He knew it was probably impossible, but he wasn't going to give up. While Phillip held Jessica, he noticed blood oozing down her arm.

"Jessica, what happened to your arm?" Phillip asked, lifting up her sleeve, noticing that she had claw marks.

"I forgot; I got scratched by the bear ... or alien ... whatever it is. It scratched me on the arm when I was trying to help Jimmy. Wow, I didn't feel



Extended Version

it before, but it hurts like crazy," Jessica replied, grabbing her arm.

"Come on; let me clean it up. The last thing we need is for it to get infected," Phillip said, helping Jessica to her feet. "Do you have some bandages and peroxide in the bathroom?" Phillip asked.

"Yes, we have some," Jessica said. As soon as she stood up, she grew tremendously weak, and before she had realized it, she passed out, and as she fell to the floor, she could see Phillip's mouth moving; he looked like he was calling her name.

# Chapter 21

The send-off


$M$eanwhile, the creature made it back to its ship. It pressed its figure against the ship, and a red circle waved around its finger; then a small drawer opened up, and the creature opened a metal band–which was on its wrist–and poured out a dozed retentballs and placed them in the drawer of its ship. Jimmy could see everything; things just looked really big to him. As he was being placed in the drawer, he yelled out, but no one could hear his cries for help. Once all the balls were in the drawer, the drawer closed back up and started processing the balls.

The ship started humming, extracting the souls that were contained in these balls, and suddenly, a white beam of light shot the souls into

 Extended Version

outer space. Jimmy saw everything moving faster than the speed of light. On the way up, he could see the spaceship growing smaller and smaller until he couldn't see it anymore; then the world started to grow smaller and smaller as he ricocheted off a satellite which was floating around the earth.

He was moving so fast through space that as he screamed, his cries echoed ten times over and over. He passed by the stars so fast that they looked as if they were flashing throughout space. Suddenly, he could see plants appear and disappear before he could get a clear look. If he could imagine how flash felt running, this was definitely the feeling.

Philip had shot past the last planet in the solar system, and before he could reconcile everything, he entered a black hole, which shot him into another dimension-where the star grew from a bright white flash to a dim bluish burst.

Extended Version 

# Replenishment

Suddenly, the velocity at which he was traveling came to a quick halt once he pierced through a cloudy orange atmosphere. He flew into some sort of chamber, a metal containment unit that was completely black in its interior. He couldn't see anything, but he could hear a gulping sound-like stuff was falling from the sky into some sort of suction tubes.

Back on earth, Jessica woke up the following morning, having passed out the night before. As she was coming to, her vision was slightly blurry. She slowly opened her eyes and noticed that she was on the couch. As she tilted her head slightly to the right, she could see Phillip asleep in the chair beside her.

For some reason, she was tremendously weak and could hardly move; her body felt as if a Mack truck had hit her. She looked over towards the door, and she could see a reflection approaching the door. She felt horrible but forced herself to her feet and

 Extended Version

wobbled over to Phillip, tapping him on the shoulder. "Phillip, wake up … someone is at the door," she whispered. This startled Phillip a little, so he jumped, looking up at her with relief in his eyes.

"Oh, it's you … Who's at the door?" Phillip asked.

"I have no idea," Jessica replied.

There was a knock on the door. "This is Officer Dean. Anyone home? I have a few questions about Officer Phillip," Officer Dean said, knocking on the door once more. "Mrs. Howell are you there?" he asked, knocking again.

Jessica was trying to be as quiet as possible. "Jimmy has a secret hiding spot in the floor; you should be able to hide there until he leaves. Help me move the table," she said, grabbing the table as they lifted it slowly and placed it to the side. She rolled up the rug, and Phillip could see a hole in one of the slacks. He placed his finger in the hole, pulled the door open slowly and was able to climb down into the floor without making a sound.

There was a lot of miscellaneous stuff down there, from guns to a change of clothes. After Phillip was snug on the floor, Jessica pulled the door back over, but not realizing the door was so heavy, it clunked as she closed it. Once it sounded off, she looked up 'cause she knew Officer Dean heard it; sure, he did. "Mrs. Howell, is that you?" Officer Dean asked as he tried to look through the door window, which he couldn't seem to see through. He did notice a shadow moving around inside. "Mrs. or Mr. Howell is anyone there?" he asked once more.

Jessica scrambled around, rolling the rug back over the door and slowly dragging the table back onto the rug. Not wanting to seem too suspicious, she decided to respond. "Hello, is someone at the door?" she asked as she walked towards the door.

 Extended Version

"Yes, it's Officer Dean. I needed to ask you a few questions about Phillip."

"OK, give me just a second," Jessica responded, as she made sure everything looked as it did before she had moved the rug. She walked over to the mirror by the door to make sure she looked presentable. She raked her fingers through her hair since she didn't have a comb; then she whipped the cool out of her eyes. Swaying her head to the right, she slung her hair out of her face. She looked back, hoping Phillip wouldn't make any noise while she reached to open the door. The door opened. "Hello. How may I help you?" she asked Officer Dean as she looked at him. She was so nervous her heart was pounding in her chest.

"Yes, I'm sorry to bother you, but have you seen Jimmy's father? He has gone missing, so if you have any information on his whereabouts, it will be highly appreciated," Officer Dean Justin said, waiting for a response.

Jessica hesitated for a moment and had to quickly come up with a response, or he would suspect something. "Umm, no. My husband was saying something about him being missing earlier, but I didn't think anything of it because he is a cop. I just figured he was busy handling police business. When was the last time you saw him?" she asked, hoping that he didn't notice that she might have been lying.

Officer Dean Justin looked at Jessica as if she knew more than she was putting off. He wasn't sure, so he just went along with it. "Well, no one has seen him. His last known whereabouts were at the library," he said, watching her as she attempted to respond.

"Oh, OK, well, do you mind me asking what he has done?" Jessica asked, letting go of the door



Extended Version

handle because she could feel herself starting to tremble from anxiety.

Officer Dean Justin knew Jessica was beating around the bush, so he asked, "Do you mind if I come in for a second?"

"Sure, you can–for just a second. I'm headed out in a few minutes, though," Jessica responded. Officer Dean Justin walked in, and Jessica led him over to the sofa. "You can have a seat here."

"Thank you. Well, technically, he hasn't done anything. His partner has gone missing, and he was the last witness on the scene, so I have to bring him in for questioning. He's the only lead we have."

"So, in other words, you think he has kidnapped or murdered his partner?"

"I didn't say that."

Phillip tried to stay as still as possible as he listened to everything that was being said. He was

Extended Version

growing impatient and started moving slowly around because he couldn't see.

"You didn't have to. I think you've overstayed your welcome. Phillip would never do something like that. As long as I've known him, he has never come off like that type of person," Jessica said as she stood to her feet slowly.

"I don't put anything past anyone, but he is innocent until proven guilty," Officer Dean Justin replied.

"You can leave now. There is nothing more we need to talk about," Jessica said, lifting her arm up towards the door.

"Thanks for your hospitality," Officer Justin replied as he started walking towards the door.

Phillip was consciously moving around and ended up kicking a plank by accident. "Shit!" he mumbled as he balled up his fist and put his head on it, hoping no one heard. It was loud enough that

 Extended Version

Officer Justin heard it, and so did Jessica, so she acted as if she was tripping over the coffee table. Officer Justin turned around to look and saw her tripping. "Be careful, Mrs. Howell," he said, looking around and listening to some other type of movement. Her tripping over her own coffee table was a little suspicious to him.

"Thanks, I will," replied Jessica, as she walked over to the door so she could close it behind Officer Justin. She waved as he walked off, and he waved back as the door completely shut.

# Chapter 22

Awaken

A chill ran down his body, as he started to come to. His body felt colder than usual. Even though it wasn't as cold as a winter in New York, his body temperature seemed off. When he came to, he noticed that he was in some sort of glass chamber that was full of a mist-type substance.

It wasn't white, as we humans know it to be it was like an orange-red cloud. The clouds swirled around in the chamber like it was attached to a ventilation system. The chamber had metal tubes, which ran every sort of way; some ran to a big circular object laced with pearl shaped bolts. It was shaped like a huge circular horn, and it hummed like a hummingbird would, a light circling around its

          Extended Version

core. The light flickered through these small holes that went around in a circle the size of a basketball. This device was about 10-foot tall and 8-foot wide.

There was a huge glass sphere that contained a blue cloud matter, which flickered like lightning was entwined. It also had a long glass tube that ran up and opened up at the top like a vase. There was a layer of water that covered the top of the vase-shaped object, which created the illusion that the room was under water.

The water was so clear one could see the sky, but instead of seeing blue skies and white clouds, one could see orange skies while red clouds floated every which direction. The glass chambers were lined up in a circle around the big glass sphere; there were about 20 chambers. The room was big, and there were tubes that hung from the ceiling, which imitated the same set up throughout the entire structure. The glass balls and chambers stretched as far as the eye could see.

The floor looked like a lake because it was covered in 12 inches of water throughout the entire place. There was some sort of minnows swimming around in this massive base of water; they were silver and covered in small orange bubbles; their mouths were like allergy eaters, and their eyes were like bees.

They sucked down all types of pollution. There were so many of them that they shined like school-fish as they shot through the water like dolphins down a coastline, shooting from one area to the next. Every time a chamber opened, they were right there, sucking down all the fluids that poured off the bodies of aliens that were being released from these chambers that were being opened accordingly.

There were several aliens riding around on some sort of hover skates; they seemed to be

 Extended Version

monitoring the chambers like some sort of species specialists, and every time an alien was released, these 'species specialists' scanned them with some sort of green rod. If the rod turned blue, it was an indication that the alien was in good health, and if it turned white, then the alien wasn't, and the floor would open up just under the alien, and it would fall into a dark hole.

There were also these big green aliens, which had lots of tentacles, with hands on the end of each tentacle. Their bodies were jagged; spiky and tough like a rhino's hide. They looked around with tremendous head movement as if their heads were going through a glitch phase. They had eyes lined up around the top of their heads so they could see every which way.

They also had one huge eye in the center of their faces. When an alien was cleared, one of their smaller eyes locked onto the alien, and they twitched their heads around to lock onto the next alien as their tentacles waved around, snatching multiple aliens at once. Their bodies glowed a dark green metallic color as they performed their duties. These aliens possessed a power that allowed them to levitate the healthy aliens and place them on a glass sidewalk-looking object that acted like some sort of conveyor belt.

The aliens were strategically placed so once the sidewalks came together, the aliens wouldn't bump into each other; instead, they would line up perfectly. The glass sidewalk lit up and glowed as the aliens glided across its surface. The sidewalk led to a reddish bright light, which ran out of the chamber area.

Finally, the time had come for Jimmy's chamber to open; the chamber started to rise slowly. The slimy substance that covered his body caused

 Extended Version

him to slide out of the chamber and into the puddle of water, and as soon as his feet hit the ground, the fish-like creature started to devour all the substance that slid down off his body into the water. They ate it up like piranhas would a piece of flesh. Once all the slime was devoured, the fish-like creature sped up to the next.

Jimmy was still in a daze and couldn't move his limbs yet. He was aware of everything that was going on, and it scared him to death. As he stood, an alien hovered over him, waving a green rod; it glitches, then turned white, then blue within a second, and the alien hovered in front of him, staring at him straight in the eyes for a couple of seconds. *Shit! Why is it standing there in front of me?* Jimmy thought as the alien's eyes glared at him. Its eyes looked like a ninja star, and as it stared at him, its eyes dilated. The alien suddenly waved the rod once more, and it turned white, so the alien sped off on the hoverboard.

Suddenly, Jimmy was lifted in the air and placed on the glowing sidewalk, which was doing something to his body. It was as if it was giving his body life, and he could feel tingles in his feet, then up to his knees, and so on. He still wasn't able to move, but the feeling was something he had never felt before.

He could feel his heart go from a dull beat to a thunderous pound once the tingle passed his chest area. About time, he had reached the bright reddish light; he was fully functional, and as he opened his eyes, he walked, stepping off the sidewalk and onto a planet-there were several planets, and they all looked alike. He looked up as he walked out, and he could see the red clouds he was also surrounded by trees.

 Extended Version

# How Deep Does It Go

The air was breathable. As he turned and looked back, he could see the tunnel that he had come from, and a tube that led down into a lake. He could see millions of fish skipping across the water like tossed pebbles. The lake stretched as far as the eye could see; the water shinned a light orange color.

Jimmy turned and looked back towards the forest, and he noticed animal shapes changing into alien form and alien shapes changing back into animals. There were hut-shaped homes in the treetops, and animals were climbing trees to reach the huts. The ground was covered with a blood-colored, clay-looking gravel. This planet was the closest resemblance to earth that Jimmy had ever seen. Everything was so similar but different at the same time.

# Replenishment

The tree's branched off in every which way; the grass was green. Birds flew, deer grazed upon the grass, monkeys swung from the tree limbs. If it weren't for different colors in the sky and all the weird shape-changing, it would be hard to tell that Jimmy wasn't on earth anymore. He was so amazed that he had almost forgotten that aliens surrounded him.

Jimmy started easing over towards some nearby bushes so he could get out of sight. "How in the hell did I get here, and where in the hell is this place? Man, I have to figure out how to get back home. I know my wife is worried sick, but I guess I have to figure out how to get off this planet first," Jimmy said as he scanned the area.

 Extended Version

# Chapter 23

The Hunt


Back on earth, it was 6 p.m., and Jessica was having a hard time dealing with her husband's disappearance, and what Phillip had told her was a little hard to accept. Not every day does someone tell you that the person you love has been taken away by beings from another planet.

This was something that she could not accept, so she wanted to talk to Jimmy's best friend Jonathan to see if he would give her some other type of option. She was in her car, headed to his house, and on her way, she felt a little strange; she felt as if something was growing inside of her. She looked down at her stomach and started to rub it gently as she drove to Jonathan's house.

Extended Version

# Replenishment

Jessica pulled up to Jonathan's home, and walking up to the door, she knocked a few times. Her third time knocking, he finally answered the door. "Yes, who's there?" he asked as he looked through the peephole.

"Yes, it's me, Jessica."

Jonathan opened up the door and looked at Jessica, and she immediately burst out into tears. "What's wrong, Jessica?" Jonathan asked as he hugged her. "Come in and take a seat. What's going on?" he asked once more.

"It's about Jimmy."

"Jimmy?" he asked. "What about Jimmy, is he OK?"

"I don't know. One minute he was there, and the next he was gone," Jessica replied, tears streaming down her face.

"What do you mean gone? Did someone take him?" Jonathan asked with an angry look on his face.

 Extended Version

"Someone, something; I don't know right now. I don't know if I'm going crazy. All I know, as he was standing in front of me, one second there was a bright light, and the next minute he was gone. I don't know what to think or what to do; that's why I'm here. Hopefully, you do," Jessica said, looking at Jonathan with hope in her eyes.

Jonathan got down on one knee and placed his hands on her arms as he looked at her right in the face. "I need you to try to remember something. If someone took him, I would find him. But I need you to try and remember whatever you can so I can try to come up with some sort of explanation."

"There was a bear, and Jimmy shot it, but that only seemed to piss it off. The bear tried to scratch Jimmy, and I tried to grab Jimmy and pull him out of the way, but it ended up scratching me instead. I fell to the ground, and before I could turn around and look, a bright light burst, blinding me, and when I was finally able to see again, the bear was gone, and Jimmy had vanished right along with it. I don't know what happened, and I can't find Jimmy anywhere," Jessica said as she cried her heart out.

"So, you think a bear took him? If that's the case, I will get my gun together, and the boys and I are going hunting. I don't understand where the light came from, but I will put together a searching party, and we will go see if we can find Jimmy. I don't know if he's dead or not, but we will see if we can find out what happened to him," Jonathan said as he got up off his knee and headed to go get his shotgun together.

Jessica didn't want to bring up what Phillip had told her because she refused to accept that he had been taken by some sort of being from another planet. She was just hoping that Phillip was wrong,



Extended Version

and Jonathan was able to find Jimmy and bring him home.

She knew that if she went around raving about her husband being taken by an alien, it wouldn't go too well. Most likely, everyone would think that she was crazy or was using the excuse of an alien taking her husband as a way to get around murdering him, which she didn't do.

The situation was crazy, and she weighed out all her other options before she started to believe Phillip's take on the situation. Not only was it impossible, but it also was just upright crazy.

Jonathan had made a few phone calls and had got a few of his friends together so they could see if they could find Jimmy. Jonathan had been friends with Jimmy since he got in the military 10 years ago, and they had become close friends.

Extended Version 

# Replenishment

They got together for holidays, went hunting together; they even served side by side in the war. He had just got out right before Jimmy was hurt. Jonathan had served 20 years in the military and was happy to retire. He had been in for so long that once was finally out; he had a hard time blending back in with society. It was like he was trapped in a ball being in the military; things were a lot different in reality.

Things weren't so direct and precise; having to be up at a certain time every day, long strenuous hours of work, accumulating scars that didn't heal all the way, and acquiring so much wear and tear all over his body, which caused him to wake up with aches and pains in his legs and back. Being able to get out the house for a little while was something that he had done in a while. The last time he got out the house was when he and Jimmy went out and had



Extended Version

a few drinks with the fellas to celebrate one of his military comrades' birthdays.

Jonathan had finally contacted a few of his buddies and told them to meet him at Jimmy's residence. He loaded his shotgun, pistol and a few knives in the back of his truck and said, "I'm ready, Jessica. Let's ride." He jumped into his truck and rode off. Jessica ran over and jumped in her car and followed him to her house.

A short time after, they pulled up and were accompanied by four of Jonathan's buddies. They all got out of their vehicles and loaded up their guns, strapped on their vests and placed extra clips and knives in the vests. They all were loading up like they were about to go to war; little did they know that night, they might just be in for the fight of their lives.

# Replenishment

Jonathan turned to Jessica, who seemed anxious to accompany them. "Jessica, sorry, you're going to have to stay here. I will leave you a walkie-talkie just in case we find him. Or if he comes back, then you can let us know. Just show me where you last saw him."

"The last place I saw him was on the side of the house," Jessica said, pointing to the right side of the house where they kept the garbage cans.

"OK, now go in the house and lock up," Jonathan said and then turned to the guys. "Grab your flashlights and let's roll, boys." They walked over to the side of the house, and Jonathan looked around and took a knee by some liquid that might have been some sort of blood. It was dried up, but he could still tell it was some type of blood. He dug his hand into the soil and grabbed a handful of soil mixed with some of the dried-up blood. He sniffed it. "Ummm, this is something strange; I've never

          Extended Version

smelled a substance like this before. It's surely something different." He looked up at the guys, then looked out into the woods and pointed his finger. "We are heading that way, guys. Let's find Jimmy if he's still out there!"

Before the party walked out of the fence and into the woods, Phillip came out of the house. "What's going on, guys? If you guys are going out to look for my son, I'm coming. I'm the only one who has seen the creature, who has taken it," Phillip said. Jonathan looked back. "Creature, what do you mean creature?" he asked Phillip.

"My partner was taken also, and I was able to see the creature for a brief second. I couldn't believe my eyes, but I do somewhat know what it looks like and what it is capable of," replied Phillip.

"So, you're telling me something, not someone, has taken Jimmy?"

Extended Version

# Replenishment

"That's exactly what I'm saying, so I'm coming because you guys will need me."

"OK, well, do you at least have a gun?" Jonathan asked, and Phillip pulled out his gun and cocked it.

"Yeah, I have that covered," Phillip said as he walked past Jonathan. He looked back and then asked, "You guys coming?" waving his flashlight. Jonathan looked at his post and signaled them to come. "Yeah, let's go." The guys walked towards the woods, and there were a few paths in the woods that Jimmy used when he was out hunting.

A few yards into the woods, Jonathan stopped and took a knee. He picked up a broken twig, which seemed broken from being stepped on. He also grabbed some soil and sniffed it. "Whatever it is, it definitely came through here. I can smell it," Jonathan said as he stood to his feet and let the soil fall through his fingers. "There is an old, abandoned

 Extended Version

shack about 200 yards into the woods, just beyond Jason's path, where old man Hicks used to live," Jonathan said, walking off into the woods.

The sun was starting to set; so the further they went into the woods, the darker it got. The day was a little windy, so the trees swayed back and forth; it was hard to hear anything that might have been lurking in the darkness.

They waved their flashlights around every which way as they grew further away from Jimmy's house. About five yards from the abandoned shack, Jonathan put up his hand, signaling everyone to halt, and he looked up ahead as he stood on a hill, looking at the trees surrounding the shack.

There were trees that were a few inches away, moving every which way, but a tree a few feet away from it wasn't moving at all–like the wind ceased to exist where the tree stood. It was the weirdest thing that Jonathan had ever seen, so he didn't know what to make of it.

"Guys, do you see that?" Jonathan asked as he stared at the tree, hoping that it would budge.

"See what?" one of his partners asked as they all looked.

"See how all the trees are moving, but when you look, the trees within about 50 feet from the shack aren't moving at all," Jonathan said, pointing in that direction.

Phillip had a flashback of how he was beside his partner as he was dying slowly, and he witnessed the same thing. When the creature was in their presence, the trees stood still; it looked as if the earth pressed pause. Suddenly, a chill ran down Phillip's spine. "I've seen this before! The same thing happened right before my partner was taken.

Everything around us came to a standstill. Somehow, I shot the creature, and I was able to see

 Extended Version

it for a brief second. It walked around just like you and me. It didn't have a mouth, but somehow, it's able to make this wind type of sound. It sounds like the wind is blowing, but nothing moves at all. It has hands and feet; its skin is a reddish orange color. It has different belt-shaped objects wrapped around its legs and arms, which it uses to carry its weapons. It stands about 7-foot tall; its eyes had a star shape in the center. It's a creature I had never seen before in my entire life.

After those brief couple of seconds, it vanished as if it had never existed. So, if you guys were looking for the creature that took my son, it's definitely here. So, from this point on, keep your eyes open, and your weapons locked and loaded."
Jonathan noticed some of the blood on a nearby leaf in the direction of the shack. "You're right, Phillip. It's surely here … somewhere …"

Extended Version

# Replenishment

Meanwhile, Officer Dean decided to make another unexpected visit to Howell's residence. When he pulled up, he noticed a few unfamiliar vehicles; he was hoping Phillip was in the facility. Some questions had to be answered, and he was the only one who could do so at the time. He was missing, and so was his partner, and he was the only one who was known to be at the scene at the time of the call.

Officer Dean walked up to see if Jimmy or Jessica was willing to cooperate to find Phillip. He knocked on the door; Jessica ran and opened up the door because she thought the guys had made it back with some answers, but to her surprise, it was Officer Dean. From the reaction he saw on her face, he could tell something was going on.

 Extended Version

# Chapter 24

Infected

$B$ack at the shack, Jonathan, Phillip, and the guys had loaded up and started casually heading into the area where the wind element ceased to exist. All of their stomachs were in knots, and sweat was running down their faces like they had caught a strenuous fever.

Their hands trembled while on the triggers; none of them except Phillip knew what to expect. The wind blew and tossed the bushes around and had everyone on edge before they fully entered the creature's domain. Every brush from the bush had the guys jittery as if the creature had run by and brushed up against them.

# Replenishment

The ground was slightly soggy, so their boots sunk in the mud as they trampled through the damp area. They stopped right where the grass swayed a couple of inches from where it ceased to do so. They turned and looked around at each other. "Everyone stay close and move quietly," Jonathan whispered as he signaled everyone to follow his lead.

They proceeded across into the wind barrier. As soon as they all made it across, it was amazing how their clothing went from swaying in the wind to hanging stiffly at their sides. It was as if they had entered another dimension because their reality was fluttered as they passed through.

They all wondered how the impossible was all of sudden possible, like they had found a gateway to another planet right in their backyard. Every step they took further towards the shack, the more terrified they got. Phillip could feel the creature's

 Extended Version

presence somehow. Not sure how, but he could feel that they were growing closer and closer.

They had all finally made it to the front door of the old beaten-down shack, and they all raised their flashlights and shined them through the front window, hoping that the light would pass across Jimmy's body. They had no luck, so Phillip said, "Guys, let's check around back. I don't think the creature would place him in the home. The creature seems more like an outside creature, so it may be lurking around somewhere in this area."

Jonathan put his index finger over his mouth, and then signaled them to head around the right side of the house. The shack had a wood fence that ran from one side of the home to the other. On the right side was the gate, and it was already open.

All the trash and mud had lathered up against the door, which had stopped it from closing. They all

Extended Version 361

entered the side entrance and preceded with caution down the side of the shack, shining their light in each window they walked past to make sure nothing was lurking on the other side of those old dusty windows.

When they got closer to the back of the shack, a pavement slab poked around the corner, which lay solidly across the backyard. It seemed like the further they got towards the back porch, the darker it got. As they looked around, their pupils had no choice but to adjust to the dark occasion. Just as they were about to proceed to the backyard, Jonathan heard a crowing sound, like an animal was devouring a piece of flesh. "Shh …" he whispered, as he timidly got ready, looking around the corner.

Jonathan put his back on the wall and looked directly across the dingy fence, which was softly illuminated by the clouded light, which hung from



Extended Version

the corner of the shack that was triggered by motion.

Obviously, something was lurking around the corner. His right eye grew bigger and bigger as he proceeded to peep around the corner, not sure what he was about to see. He started to slowly turn his head around, and his eyes scaled around the backyard, noticing all the scrubs and twigs that hung from the raggedy fence post, which scaled around the backyard.

Finally, having turned all the way around, Jonathan's right eye was looking at the back porch as he eased his head further out and noticed a deformed-looking leg attached to a leopard. His heart started beating hard in his chest as he looked at the leopard, wondering why one of its legs looked like it was a totally different species.

Extended Version

# Replenishment

He took a step closer to the corner and didn't realize that a hollow twig was about to share the same corner as his boot as he eased out further, and his boot smashed the twig, causing the leopard to flinch and turn around. It looked at Jonathan right in the face, and suddenly, the leopard's face flickered like a TV set with a glitch, and its face went from a leopard's face to its alien form, then back to a leopard as it heisted at Jonathan. This frightened Jonathan so badly he jumped back around the corner and couldn't stop shaking, "It's … it's … Over there," he stammered, pointing towards the corner as his finger shook out of control.

Phillip signaled to the guys to raise their weapons and keep an eye out as they walked around the corner. They all had weapons locked and loaded, ready to take down the creature. They all stepped around the corner in sequence with weapons up. They saw the creature, and before they started to

 Extended Version

fire, the creature had fled around the other side of the house. Once the smoke cleared, the guys realized that they had missed the creature.

"It looked like it was infected. Did you guys see that one of its legs wasn't in leopard shape?" Phillip asked, looking back at the guys.

"Man, you didn't tell us this creature could transform into different animals," replied one of the guys.

"Well, I technically didn't know. I can tell what it was because its leg looked like the alien that I had seen before," replied Phillip.

"This is some weird shit. We are in the woods with something that can be anything animal and could be anywhere. I don't think our odds are too high out here," Jonathan said, finally pulling himself together.

# Replenishment

"Well, we need to stick together. That thing is the only way we can find out what happened to Jimmy and if he is still alive," Phillip said.

Once the alien got on the side of the house, there was not an exit, so it transformed back into its alien shape and activated its flexure band, which let off a bright white burst of light and bent a portal on the fence through which the alien exited to the other side of the fence, and once the alien had made it through, the passage was sealed behind it.

"Look, a light just went off on the side of the house. Come on, guys; let's get that son of a bitch before it gets away," Phillip said as he raised his pistol.

The guys started moving forward, looking down at the chewed-up deer carcass that was torn to shreds, in a puddle of blood, lay across the back porch.

 Extended Version

## How Deep Does It Go

Back at Jimmy's house, Detective Dean was asking Jessica a few questions when he heard guns going off out in the woods. He looked at Jessica, then turned and ran to his car, grabbed his flashlight and sprinted out into the woods, pulling out his pistol as he proceeded off into the woods. He had a hunch that he would find some of the answers that he was looking for, with Phillip and his missing partner.

Back at the shack, the guys had lined up and jumped around the corner, thinking that they would have the creature trapped, but to their surprise, when they jumped around the corner, with guns and flashlights pointed, nothing was there. They waved their flashlights around. "Where did it go? There is no hole or anywhere it could have exited. Maybe it jumped the fence. Come on, guys. Let's get it. We can't let it get away," said Phillip, waving the guys to follow him as he ran back around to the other side where they had entered.

All the guys ran behind Phillip. As soon as they made it back to the front of the house, they noticed that the elements started to come back to life, the wind started running through the trees, and the dirt gusted up through the grass.

"It is definitely on the run. Stop for a second, guys," said Jonathan as he looked up to see where the trees weren't moving because that's the way they would head. He looked to the right and turned his head to the left and noticed the treetops to the west were not moving at all, about 50 yards out. "It looks like he went that way. Come on, guys, before it gets away," said Jonathan as he ran towards the stiff trees.

"I'm going to kill that fucker if it took my son. It already took my partner, and this can't keep happening. We have to put a stop to this," said Phillip, who then started to run off with the guys.

Meanwhile, the alien had made it back to its ship. The ship was in a small field; the creature powered it up, then placed its hand on the ship, and the ship started scanning its hand. A few seconds after, the ship started to glow with a dim green light.

Then it started opening up, the sphere started to open up, and the seat was visible as it opened. It opened downwards so the alien could step in. The step looked like glass, and one could see the grass under the door. On each side of the sphere was the place where the alien could place its hand face down.

On the hand panel, there were differently lit buttons where the creature's fingers would lay, with which the alien could control the sphere and weapons system. There were red lights around the exterior border of the seat; there was a helmet-type object where the creature's headrest was, which had some sort of oval eye cover.

The guys had gotten close enough that they could see the ship opened up and the alien getting ready to step in. Phillip pulled and aimed his gun at the alien and fired off a shot, hitting the creature right in the shoulder. The creature screeched a horrible sound that echoed through the woods. It looked around and placed its hand in the ship, arming its weapons system. Suddenly, the ship turned red and started rapidly shooting laser beams.

One of the beams blew a hole through one of the guy's head; a few of the other beams busted through a couple of trees right next to the other guys. Phillip and Jonathan hit the ground. One of the other guys screamed as their partner's body hit the ground; he turned to run away, but before he could pick up the speed to run away, the ship powered up

 Extended Version

laser had turned from red to a bright white, and the ship started shooting off into the sky. It looked like a shooting star; the guys had finally stood to their feet and watched the ship fade away in the distance.

Dean had finally made it to the area, and he could see something shooting across the sky like a star, but he had no idea that it was a ship. As he walked up behind them, he noticed the bodies lying on the ground, blown to smithereens. It was a sight that he had never seen before. He noticed that one of the guys was Phillip when he grew close enough to identify him. He raised his weapon. "Freeze! Don't any of you move! Toss your weapons down, place your hands on your head and get on the ground slowly," Dean said, standing back far enough, for he had no idea what was going on.

The guys tossed their weapons aside, put their hands on their head and slowly fell to their knees as

Extended Version

# Replenishment

Phillip looked up at the sky. He wasn't worried about getting caught anymore; his main focus was this creature that had taken his son and how he could get him back. The look in his eyes said a million words, but only one stood out at that moment-vengeance.

# Chapter 25

Proliferation Begins

$B$ack on planet Zara, Jimmy didn't realize that he had fallen asleep behind the bush that he was hidden. When he opened his eyes and looked around, the sky was a darkish orange color, and the planet was a little darker than he had remembered.

There was no activity going on; it looked like nothing existed there. There were no aliens in sight; it was if he had woken up from a bad dream. He knew it wasn't a dream because he was still stuck on this strange planet, with no way to leave and no sense of where he was even located. He had been trained to overcome difficult situations, but this wasn't something that he was trained for. The only positive thing was that he was alive and somehow survived traveling through the deep corners in space.

He was still not sure how he was able to breathe on this strange planet when there was no proof of life, let alone oxygen that could sustain these life forms. Jimmy was intrigued that this was even possible; plants were growing, and there was a lake full of water and clouds in the sky. The look wasn't one like the earth, but it was close.

As jimmy looked around, he could see something cracking through the atmosphere, traveling at an uncontrollable, tremendous speed. Whatever it was, the object was covered in a bluish fire as it screamed through the sky on a crash course towards the east side of the lake.

Jimmy suddenly had a flashback of the war: As he looked up, he could hear a missile headed right for him. He jumped up and leaped; it hit a couple of yards away, and the force of the impact pushed



Extended Version

Jimmy so hard that his body flew uncontrollably right toward a motorcycle that had been demolished.

Suddenly, Jimmy came out of the flashback, breathing hard because it seemed so real; he thought he was right back in the middle of the war. It was so vivid; it freaked him out for a second, his heart pounded in his chest, like a thump from a base drum in a drum line. After a couple of seconds, he was able to shake it off, and he started looking up again.

He didn't have a clue on what it was, but he was hoping that it would be something that could benefit him in some sort of way, so he started to run towards the falling object. As he headed on the course towards the object, he could hear it smash into the earth. He looked up and could see a cloud of red dust, which kicked up as the object crashed into the planet.

The closer Jimmy got, he could hear wind sound whistling about 50 yards away from him, close to the lake. Moving even closer, he could see that the lake had opened up on the west side; the tunnel led down under the lake. So, he figured that he wasn't the only one who had heard the object smash into the planet.

Jimmy ran, but he stayed close to the bushes as he approached the scene.  For some reason, Jimmy was moving really fast and didn't know why he had transformed into a cheetah and was running faster than he could ever imagine.

Even though he had transformed into a cheetah, he had no idea that he had transformed, let alone that he was in an alien body and had the ability to transform into anything that he had a thought in his mind about. He had gotten close

          Extended Version

enough that he had a clear sight of the crash and could see about 5 aliens standing at the crash site as he approached.

He went from running on all fours to standing in his original form. He could see that one of their ships had crashed, but it looked perfectly fine, like it had some sort of crash protection system in place. One of the aliens placed its hand on the ship, and it started to turn upright, then the door opened up. A light mist hazed out, as an alien lay inside, looking as if it was dead. Once the door had completely opened up, its left arm had fallen out; the creature was dead or hurt really bad.

Suddenly, the aliens started to pull the corpse from the ship. One of the aliens dropped a small silver ball on the ground, and it transformed into a floating table that they laid the corpse on, and it hovered in the middle of the group as they walked

Extended Version

away. Once they had gotten a couple of feet away from the craft, the door closed back up slowly. Jimmy watched the aliens fade off into the distance. He was still a little curious about the ship and was thinking that this could be his way back home.

Jimmy walked up to the ship and looked at it, and a soft white light blinked slowly around the center of the ship. Jimmy slowly raised his hand to touch the ship. As he touched the ship, it scanned his hand, and he noticed that his hand was not his own.

The ship started to open up; Jimmy raised his hands-it frightened him so badly that he jumped and fell backward onto his butt, and at that moment, he noticed that he wasn't in his body anymore. He looked at his arms, hands, legs, and feet. He was definitely not in his own body anymore; somehow, he had been placed in the body of one of these creatures. His skin was a pale orange color; the features were similar to the human body.

 Extended Version

Things like fingernails and toenails were not there; his skin looked slightly scaly but was smooth like human skin. He slowly raised his hand to his face and started to feel around–his nose was gone, and he had no mouth. His heart thumped. "NOOOOOOOOOO!" he yelled, which sounded like a loud stretching sound, almost as he was hurt. His cry echoed so loud that the aliens heard it. They had just made it back and placed the corpse on an escalator that lowered it down into the main quarters.

The aliens immediately turned and started to run back towards the ship. Jimmy could hear them approaching, so he jumped up and walked towards the ship. He looked at it, and he noticed some symbols on the ship that he didn't recognize.

He closed his eyes and focused and looked again, and everything was clear as day, and he could read everything on the ship. He climbed in and

Extended Version 

looked for something labeled 'weapons.' He saw a button that would trigger the weapons system.

The aliens had made it to the ship and stopped and looked; suddenly, they started to run towards Jimmy with extreme prejudice. He triggered the ship, and something popped up, saying, "Humans or All." He hit the 'All' option, and suddenly, the ship turned a purplish color. The creatures knew things were about to get worse, so they started to transform into different animals.

One transformed into a lion, the other into a panther, another a black bear, and the last a cheetah. Before they could get close enough, the ship started shooting purple bust-shaped bullets at each one of them, and they started dropping like flies. One of the bullets denigrated the lion's head, another blew off the bear's leg, another caught the panther right in the chest, and the last couple blew a hole through the cheetah the size of a bowling ball. As they died

 Extended Version

and hit the ground, they morphed back into their original alien form as a purple smoke steamed from their corpses.

Jimmy was amazed. "Oh shit," he said. He had never seen anything like this before in his life. It had taken out all four of them within a few seconds. "Wow, this is one of the craziest weapons systems that I've ever seen," he said.

Jimmy had been in the armed services for a while, and some of the most advanced weapons had nothing on this one small ship that he was in. The ship could take down a few jets in a matter of seconds. There was an ignition switch, so he could take flight, and a map that was programmed, which would take him, right to the earth.

He was so happy and had forgotten that he was in an alien body or that there was a system in play that would steal human souls to bring aliens on this planet back to life. Just as he was about to push the button, a million thoughts ran through his head. *Could I stop them? How many others here, like me, are trapped in these alien bodies? Are we the reason why this alien species exists? How many aliens are on earth? Can they transform into anything?* All these questions were rushing through his head as he thought about pushing the button.

Suddenly, Jimmy jumped up and climbed out of the ship. "I can do this. I have to destroy the machine that captures the souls when they make it to our planet," Jimmy said as he walked past the corpses, looking down at the bodies. He noticed that the corpses had not fully transformed back into their original forms.

They all had at least one limb that was still a part of the animal's body that they had transformed into. It was almost like something was wrong with them, a dysfunction of some sort. It looked

 Extended Version

disturbing to him-to see the alien corpses disfigure-but he had a mission now, and he was determined not to let his planet down.

He was walking past an alien when he noticed a band on its wrist, so he took it and placed it on his arm. He wasn't sure what it did, but he was sure he would figure it out sooner or later. He also noticed that they had a band on their leg, with a pouch attached to it. He thought, *this must be something that holds some weapons.* He removed that, too, and placed it on his leg, and once he did, the two bands beeped as if they were in sync; then they tightened down around his wrist and leg. This frightened Jimmy a little, so he jumped. The band around his wrist suddenly lit up, and codes displayed on it; somehow, Jimmy was able to read it. There were a few words he made out:

Flexure Truss

Extended Version

# Replenishment

Snatcher More

Retention Ball

Deaddimizer

Readdimizer

Jimmy had no clue what any of this meant, so he left it alone for the time being. He got back on track and started to get his mind right for the journey ahead of him. His military training was something that would definitely help him in this situation.

The military had certainly instilled the proper thought process when it came to difficult situations, like the one he was about to face. Most people would have just climbed back in the ship and left, but something inside of Jimmy wouldn't let him do so.

Not giving up was something that was instilled in him as a little boy. His father always taught him to never give up, for anything was possible with the right strategy. He could remember his father telling

     Extended Version

him, "If something can be thought up or created, then it can be solved or destroyed." This stuck with Jimmy his whole life and was a vital piece that held him together when he was in the war. So, just like this creature created this machine, it can be disabled in some sort of way; the only thing Jimmy had to do was figure out how to disable their creation.


****Flashback****

When Jimmy got out of the machine and was standing there, his vision came to him, and he noticed that all the capsules were connected to the main tube that ran down into the ground.

**** ****

Jimmy wondered if he could find the secret passageway that probably resided somewhere a few yards away from the main laboratory. He thought it would probably be a tunnel that was in plain sight, for they had no enemies on their planet, so it was

Extended Version

needless to try and hide it. So, Jimmy began to scan the area a few yards away from the lake. He started walking around the lake, looking for some sort of entrance. He headed counterclockwise to the lake to start his search.



Extended Version

# Chapter 26

Interrogation

Phillip was sitting in the interrogation room, waiting on Officer Dean to come in and start questioning him. He was worried that his partner had been taken or got arrested. All he could think about was the alien who had captured his son on a course back to its original planet, and there was nothing he could do about it.

Most say that they know what it feels like to lose someone, but a child is something totally different. Not even he knew how he truly felt; he just knew that he possibly would never see his son again. Just standing there, watching the last chance to see his son fly away, had him in a state of shock.

# Replenishment

At this point, he didn't care if they blamed him for his partner's disappearance or the fact that there were bodies blown to smithereens in the woods. This was probably the worst thing that ever happened to him in his entire life.

First, he lost his wife a while back from gambling their money away; now, he lost his only son to an alien abduction. At this point in his life, there's nothing any man can take from him that he would actually care about. Them stripping his freedom away wouldn't do anything but give him time to soak in his sympathy.

While he was spaced out, the door slowly started to open, and Officer Dean walked in. Phillip didn't pay him any attention, for his mind was elsewhere. Dean stood there for a second, looked at him as he stared at the wall.

The look in his eye was somewhat detrimental, and Dean wanted to really know what

 Extended Version

happened. He couldn't find Wayne's body, and Phillip was the last person to know his whereabouts. The only thing that puzzled him was why Phillip fled the scene. At this point, Wayne was missing, and there were a few guys with holes blown through their bodies, and he knew this wasn't something done by any regular gun. Dean walked over to him slowly, pulled out his chair and took a seat.

"Phillip, I need you to tell me what's going on. Your partner is missing, and you have some dead bodies on your hands. You were the only person known to be at the scene before your partner went missing. You also were just caught at the scene with a few dead bodies. I want to help you, but I need to know what happened. Let's start with your partner; where is he? What happened to him, Phillip?" Officer Dean Justin asked. He looked at Phillip, waiting on him to say something. "Phillip! Phillip!"

Extended Version

# Replenishment

"Yeah" Phillip responded as he slowly turned and looked at Officer Dean.

"Did you hear me? What happened to your partner?" asked Dean once more.

"I … I … I don't know," Phillip replied as his bottom lip shivered.

"What do you mean you don't know? You were the last person at the scene. I know you know something. Just take it from the top and tell me what happened," Dean said, looking at him with curiosity.

**** Flashback ****

"OK, we went out to the hunter's spot that the missing hunter was found, and we ended up going into the woods to look for some sort of clues. We were out there before and found some tire prints with no truck. The truck had driven out, but it was gone with no marks to show that it had left.

 Extended Version

We went into the woods and ended up finding a piece of one of their rifles, which was melted down. So we were out in the woods that day, looking for more clues. It was weird, though; none of the trees were moving once we entered the woods. That day, we went out into the woods, we noticed that the trees were moving all around us, and about 25 feet in front of us, the trees were not moving at all. I was down on one knee, looking at the spot where we left off. Before I could get up, Wayne had started walking towards the area.

When I noticed that he was almost there, I got up and started heading that way. Once I had gotten about 15 feet from him, a gold beam of light busted through a tree and pierced a hole right through him. I was able to make it to him and pull him to safety. I told him to hold on, and I was able to get him back to the car. I called for backup, but it was too late …

Extended Version 

# Replenishment

I thought that was the end; then something came out of the woods behind us.

Whatever it was, it took him. I was able to get a shot off, and I hit it. Suddenly, there was a black ball, then a bright light, and when I was able to see again, he was gone. Once that happened, I got scared and ran off. I didn't know what else to do. I didn't know what to think or do. I know you're probably not going to believe me, but that's what happened," Phillip said as his hand trembled on the table.

*** ***

"OK, sir, I did believe some of what you said. Who took him? Who did you shoot? We found his blood but no body, so that has us a little lost. You are saying you don't know where he is and that something took him, so what took him?" asked Dean, trying to be reasonable.

"I'm not sure what it was. They call them 'Whiffers.'"

          Extended Version

"Whiffers? What in the hell is a Whiffer?"

"I'm not sure what they are or where they come from, but they are what's taking all these missing people."

Dean placed his hand over his face and took a deep breath. "Phillip, I'm trying to reason with you, but you are going on about something called a Whiffer. If you were me, what would you do? You are giving me much to work with. The only thing I can do is go back and check out the area where you're saying something shot him and see if his blood is on the tree that you claim–where he got shot. If you waste my time, and I go back out there and find nothing, I will put you under the jail and throw away the key because, for some reason, I think you are wasting my time. If you killed your partner, just tell me so we can end this. Just tell me how you killed him and where you put his body so

Extended Version                      395

his family can try to get some closure." Dean hoped Phillip would just confess so they could solve the case.

"I didn't kill my partner; you have known me long enough to know Wayne was like family to me. He has been the only one there for me since my wife left me. Dean, I did not kill my partner," Phillip said, looking at Dean directly in the eyes.

"OK, this is some bullshit but. OK. Next, what happened to those guys in the woods? They were definitely not shot by any weapon that we found on you guys. So, what happened to them? And please don't tell me the Whiffer shot them," Dean said with a look of frustration on his face.

"Well, it's the truth. I don't know what else you want to hear. There are unsolved cases of missing people that happen every 25 years. Cases that you or I have never solved matter-of-factly; there are no records of anyone solving cases when people

 Extended Version

disappear during the 25th year mark. So, you can keep calling me a liar and go do your job and find out what's actually going on.

You know there is something serious going on out there, and it's not me. The end of the month is growing near, so trust me when I tell you more people are going be to go missing before it's all said and done. So, if you're going to lock me up, then go ahead because I've told you all that I know," said Phillip.

"I don't know what type of master plan you and your little partner next door have going on, but you guys really worked hard to make sure your stories matched up. I don't know if you both are insane or trying to become some serial killers.

It's cool, though, because both of y'all will be locked up until we gather something so we can lock you both up for a long time. So, while you're locked down, think long and hard if this is what you really want," Dean said, looking agitated from the conversation they were sharing.

Phillip looked at him right in the eyes. "Dean, they took my son; that's why we were out in the woods. We were trying to catch the creature. It killed those guys and got away right before you got to us. That was the only chance I had to find my son, but I guess it's over now. The creature left, and there was nothing I could do to stop it," Phillip said as tears ran down his cheeks.

"So, Jimmy is missing; are you serious?" Dean asked, looking a little ashamed.

"Yes, he is. I didn't find out until Jessica told me what happened. So, we got a group together to go see if we could find him before it was too late. You have seen how the story ended, though," Phillip replied, putting his head down, trying to hold in his emotions.

 Extended Version

"Hopefully, he just got lost or something. We will set up a search party and see if we can locate him. We still have to lock you down for now, so sit tight, and we will get you back to your cell, and we will get back to work," Dean said, looking at Phillip, feeling a little pity.

Dean walked out, and another officer came into the room and moved Phillip back to his cell. He was in a cell right next to Jonathan. As he walked by, he could see Jonathan sitting there like he was still in shock. This was a lot for both of them to have seen.

Jonathan lost some of his best friends, and he had lost his partner and his one and only son. Phillip had never seen weapons that could blow a hole through human flesh the way it did. Seeing those men get shot down like a duck during hunting

Extended Version

season was definitely something he would never forget. This was a dramatic experience, and not having anyone that he could talk to that would actually believe him was very frustrating. Officer Dean looked at him like he was crazy when he was explaining what happened.

Phillip looked over at Jonathan. "Hey, man, how are you holding up?" he asked.

Jonathan sat there for a second before he even realized that Phillip was speaking to him. "I don't know how to feel right now. I just witnessed something that is impossible or, I should say, unbelievable 'cause, obviously, it is possible because it just happened. What I don't understand is how and where did these creatures come from? Is there even a way that we can fight against them? There are so many things going through my head, and its weaponry is on a whole other level compared to the stuff we have.

 Extended Version

# How Deep Does It Go

The way things are looking; we might not even have a chance against these things. It was five of us, and we didn't have a chance against one, so what if there were hundreds? And if they are, then they could come whip out human existence easily.  All I can do is keep thinking about how that ship blew them away with ease. It just keeps playing over and over in my head, and it's driving me crazy that we couldn't do anything to help. All we could do was watch as my friends got taken down one by one. I want to kill that thing so bad," Jonathan said in an angry tone as he slammed his balled-up fist into the metal bench that he was sitting on.

"I know how you feel, Jonathan. Seeing that ship fly away and knowing that I won't see my son ever again is killing me inside. I'm so angry, and I feel so helpless at the same time. We walked right into our worst nightmare. Seeing someone that we care about being taken away from us and not being able to do anything about it …" Phillip said and then took a deep breath. "I do know that if and when they do come back, I will be ready, and they will feel my wrath. They took my son, but I guarantee that they will not take anyone else that I care about," Phillip said, clutching his fist tight.

"You are right about that. Nobody believes us right now, but they will when we dump an alien body on their front porch. They might not know it, but those sons of bitches have declared war," said Jonathan.

"Yes, they have. We will be ready for them next time," Phillip replied, staring off into the distance.

Meanwhile, Officer Dean had set up a search party to look through the woods so they could try and locate Jimmy. They had decided to start in the

     Extended Version

woods where he had found Phillip and Jonathan. Dean was really hoping that they find him or at least find something that would suggest that Phillip and Jonathan could be innocent.

In his heart, he knew Phillip was a good detective. From day one, Officer Dean envied him and wanted his job, but this was not the way that he wanted to get it. He believed in working hard to accomplish his goal, not for the simple fact that he would receive it because he would be incarcerated.

Officer Dean stood at the edge of the woods and just stared off into the woods. "I hope something comes from this. I hope we find Jimmy," he said as he took his first step into the woods.

# Chapter 27

Underway

Jimmy had been scouting the planet for a while and had finally come across a tunnel that was out of the ordinary. There were no bushes to hide it or anything to make it unnoticeable, so he figured that he might have seen what he was looking for. He decided to survey the area for a while to see if anything out of the ordinary would go on. He wanted to see if any of the creatures would go down into the tunnel or if any of them would come out.

He found a bush to hide behind a few yards away so he could keep an eye out. He had gotten comfortable, and before he knew it, about an hour had gone by, and just when he was about to leave and head down into the tunnel to take a look around, he saw a caribou walk up to the hole; then it looked

 Extended Version

around and walked down into the hole. Its back leg was deformed like the other aliens earlier; it was almost as if it was sick.

A few minutes later, it came back out of the hole as if it was running away from something. Before Jimmy knew it, the ground started to quake, as something unground was moving around.

Suddenly, the ground started to split, and then something big came from the ground–the ground busted open, and a big alien came from its depths. Red dust clouded the skies surrounding the creature, so Jimmy couldn't get a good look. All he could see was a big body, and it transformed and shot off into the sky.

When he looked up, it looked like some sort of dinosaur, almost looked like a pterodactyl, but he didn't think the aliens knew about dinosaurs, so how would it be able to transform into such a massive beast? The beast's wings had so much power that it blew the dust right towards blinding him; it took off, headed right towards the animal that fled from its chambers. When Jimmy could see again, the beast had faded off into the distance.

"I don't know if my eyes are playing tricks on me, but that really looked like a fucking pterodactyl. Wow! I don't know what to think right now." Jimmy headed over towards the hole, which was ripped wide from the creature exiting. He approached the hole with extreme caution, looking off into the distance, trying to make sure the creature wasn't coming back. "Something definitely has to be down here in the pitch-black hole." He laughed and shook his head as he climbed down into the hole.

Once he got down into the hole, he noticed that it opened up into a big tunnel and went straight down into the planet. Surprisingly, his eyes adapted to the dark elements, and he was able to see as



Extended Version

clearly as if he was walking around on the surface. It was amazing how he was able to adapt to these circumstances. The further he got down into the hole, he noticed that there were these boxed shaped units about 30 meters away from each other, which were lined up, down the tunnel, and they were connected by some sort of pipes.

The pipes had lights, which flashed from one unit to the next all the way down the tunnel constantly. He had been walking for several minutes and had finally made it deep enough that he could see that the tunnel had started opening up into a big sphere area, and as he looked around, he noticed the walls were shining like diamonds, and the more he looked, the more the walls looked as if they were slightly moving; the walls were contracting as if it were breathing.

Extended Version

# Replenishment

There was a big glass bowl that was about 20 meters wide in a circular dimension; it looked like some sort of huge ball was supposed to be placed there. All the pipes from the walls of the tunnels were connected to the outside parts of the bowl area, connecting to all sides. *Whatever goes in that spot must be the source of the planet's power,* Jimmy thought, for the pipes that connected were huge. Jimmy was so curious that he walked up and placed his hand on the bowl, and as soon as he did, the walls moved like a wave had run across it.

Jimmy's touch triggered something; the pterodactyl was ripping the animal to shreds but felt a kick in its body from Jimmy touching the bowl. It turned around and looked back towards its hole. Suddenly, the beast took flight and started heading back towards its layer, leaving the corpse that it had demolished right there.

 Extended Version

## How Deep Does It Go

Back at its layer, Jimmy had taken his hand off the glass, and the wall started to move. There were roach-type creatures emerging from the walls and climbing down to the ground, headed right for the tunnel exit. The creatures' backs lit up and sparkled from their diamond-plated shells as they fled up the tunnel. As they moved, it looked like water was sparkling and rushing upwards out the tunnel.

Jimmy felt like something bad was about to happen. He was running out of options, so he decided to find a way to destroy the glass bowl or clog up the pipes so it wouldn't work properly. He had to think quickly; he started looking around, trying to see what he could do. He noticed that it had grown really quiet all of a sudden.

All the roach-type creatures had fled out of the tunnel. He knew that if he wanted to do something, then he had to do it quick. Before he could come up with a plan, there could be something large coming back down the tunnel, with the millions of footsteps from the smaller creatures.

He panicked and started looking for an escape plan. He looked around and saw that there was a wall, which could possibly have another way out. It was hidden pretty well, but he noticed it just in time. Jimmy hurried towards it and looked behind it, and there was another tunnel, which could possibly be another way out.

Just as he walked behind the wall, the huge creature had made it back into its chamber. Jimmy peeped through the wall and saw the massive creature.

"Wow, that must be the queen!"

It stood up six legs and had a huge abdomen; the spiracle was a bright orange color. The roach-type creature carried it on their backs as it prowled around on its stalky legs. Its head had vessels, which

 Extended Version

cascaded from the back of its head like plates which ran down into its abdomen. It had eyes, which were in a V-shape, which started off tiny and got bigger, then smaller as they reached the top of its dark red skull. The queen could move around very quickly, and the creatures that carried its tail didn't miss a beat either; they stayed with the queen every inch of the way.

Jimmy didn't want to get spotted, so he hurried down the tunnel after looking at the queen for a brief second. He made it down the hall, and it opened up into another big dome. Mist grazed around a slight hill in the middle of the area.

There was a huge jagged reddish egg, which sat on the top of the hill, and as soon as he stepped into the mist, he could see thousands of smaller orange and red fussed eggs that were lined up

Extended Version

around the hill. The smaller eggs weren't too small, though.

He was able to stand behind one of them and not be visible. Before he knew it, the queen was scurrying up the tunnel right towards him. He ran into the layer of eggs, not realizing that the eggs were laced in a small puddle of water.

"If there is water in here, then there must be a way that it is getting in," Jimmy said. He immediately thought that if he could find where the water was coming in, then perhaps he could open up the hole so it could pour in and flood the chamber and everything down there. There may be a slim chance that he could drown as well, but he had to take the chance.

The queen entered the chamber and started stepping over and around its eggs, trying to find the intruder. It moved swiftly around, making sure it didn't graze any of the eggs. Jimmy proceeded

 Extended Version

deeper and deeper into the chamber of eggs and constantly scanned around, looking for where the water was coming in. The queen was growing closer and closer to him; he was cutting it close and was about to be discovered.

After he had made it almost to the back of the chamber, he could see where the water was spilling in. There were a few rocks that he might possibly move so the water could get in at a faster pace. Before he could make it over to the area, the queen was right over him, and he didn't realize it until it placed its leg right beside him while he hid behind one of the eggs.

The queen stood tall, so he was right under it. It went to look down, and before it saw him, he inched around to the side of the egg, out of its line of sight. It slowly looked around and figured that the intruder wasn't in its egg chamber, so it started to walk back out slowly.

When the queen moved far enough away, Jimmy started to make his way towards where the water was coming in. On his way towards the opening, he bumped into one of the eggs, and just as the queen was exiting the chamber, it stopped and looked back, and it saw Jimmy making his way towards the water base, and it chirped so loud that the walls trembled. Jimmy turned around and locked eyes with it; it made him realize it was now or never.

Jimmy took off towards the water entrance, and the queen scrambled after him. He made it and started to pull the rock from the hole. His heart started pounding in his chest. He placed his hand on a big rock, and it wouldn't budge; the queen was growing terribly close. Before he knew it, he transformed into a bear and was able to budge the rock out of the way. As soon as the queen reached

 Extended Version

out to stick its leg into his back, the water bust in, pushing Jimmy out of the way of its vicious attack.

The water was coming from the lake; so all the pressure consumed most of the chamber, washing everything out of the chamber, including the eggs. Jimmy and the queen were submerged under the water and were scrambling for their lives.

Jimmy thought about swimming and was able to transform into a dolphin and started to swim out of the chamber, up through the tunnel and little did he know the queen transformed into a shark and was swimming slowly behind him. A few minutes later, he was able to make it out of the tunnel, back onto land. He climbed out of the hole and lay on the ground, gasping for air. Once he got his breath up, he was able to make it to his feet, and he started running back towards the ship so he could get off this dreaded planet.

Extended Version

# Replenishment

The queen had made it to the surface, just as Jimmy started running off. The water had drained the queen's energy, so it wasn't able to transform again, so it had to pursue Jimmy on its feet. The only problem was some of the creatures that carried the queen's abdomen drowned in the flood. About 20 of them luckily made it out, so the queen wasn't able to move as quickly and swiftly as before.

Jimmy was almost back to the ship and looked around and noticed that the lake was about half empty, and he could see the underwater alien laboratory, and it was smoking and falling apart. He had destroyed everything and ruined their planet. He was so happy that he started to celebrate, not realizing that he was being followed. The queen was inching up on him; it couldn't move as fast as usual, but it was making decent strives.

While Jimmy was in celebration mood, the queen was able to sneak right up on him; it

 Extended Version

screeched and slammed its claw down right at Jimmy. It startled him, and he fell backward on it, but just missing its blow. He looked up, and he was looking at it straight in the eyes.

He started slowly inching back. Suddenly, he jumped to his feet and started running full speed towards the ship, avoiding every striking blow that the queen threw at him as he tried to run away. It was so big that Jimmy couldn't seem to gain enough space between them to escape. He was so scared that his emotions triggered something in the band that he was wearing, and something flashed across the display, flashing in red.

Snatcher More

Jimmy was trying to hold his balance as he dogged the queen's attacking attempts. He jumped

Extended Version                    417

out of the way, and while he was in the air, he turned towards the queen and hit the button on his watch.

A white ball shot in the queen's direction and exploded, but whatever the effects were didn't affect the queen, but the creatures that were holding the queen's abdomen froze right in their places and caused the queen to fall flat on its face. Jimmy hurried up and jumped to his feet and took off running towards the ship; he was about 10 yards away.

All he had to do was make it to the ship so he could trigger the weapons system so he could kill the queen. Jimmy was running full speed and was about 5 yards out when the queen was able to drag its abdomen close enough that it was able to stretch out its leg and trip Jimmy, causing him to stumble and roll.  His wristband lit up again:

Stun Gun.

 Extended Version

Jimmy hit it again, aimed his arm at the queen, and a weapon formed, covering his hand, then a triple short beam bust of silver light shot 3 shots right at the queen. He hit it right in the eye and caused it to slinger its head around as it screeched even louder.

Suddenly, Jimmy looked back and could see hundreds of aliens transforming into all sorts of different animals with deformed limbs headed right towards him. He was able to get back to his feet and get back to the ship just in time. He sat down and triggered the weapons system, and the ship powered up and started unloading at anything that moved, blowing limbs off in every which direction.

The queen was so big that the ship was unloading on it all over, blowing off its legs, blowing holes through its body and decapitating its abdomen, causing it to explode, gushing orange ooze every which way. Jimmy hit autopilot, which had the earth location programmed in, so the ship started powering up, as it started hovering away from the planet.

Jimmy had reached a high enough altitude that the ship was out of distance of the aliens, so the weapons system stopped firing. Just as the ship was about to pick up speed to take off into space, the queen had gotten up enough straight to transform into the pterodactyl again and was flying right towards him. The ship quickly powered up, just as the pterodactyl was about to collide, and shot a huge beam bust and blew a hole in its head; it fell right out of the sky and hit the ground so hard that all Jimmy was able to see was a big cloud of red dust kicked up as he flew away into space. He was finally heading home, and the queen was destroyed, so it was all over.

 Extended Version

# Chapter 28

Doctor's visit

Jessica and her mother, Deborah, were at the hospital because Jessica had been feeling pretty bad for the last few weeks. She was going through a lot; she lost her husband, her father-in-law was locked up, she kept having morning sickness, and she couldn't seem to hold food down.

She was constantly in the bathroom, throwing up as quickly as she could devour it. She thought she might have a stomach virus because the last time she felt like this, Jimmy had left for the war. It lasted for a few weeks, and with a little medicine, she got better. This time was slightly different; it had been about three weeks, and her home remedies weren't working this time around.

Extended Version

# Replenishment

Her mother taught her a home remedy that she had been taking since she was a little girl, and she had continued in her household. Some tea, a little honey, and a shot of whiskey usually did the job. Her mother had been around since Jimmy went missing. The house just hadn't been the same since he'd been gone.

Jessica came home and just sat in the living room, looking out the window for hours, hoping that one day he would walk up and open the door so she could run up and hug him and kiss all over him. After she got tired of sitting in the front room, she went upstairs and pulled her shoebox from under her bed, which was filled with letters that Jimmy sent her when he was away. She kept every single last one of them, and her face light up every time she opened one up and started to read it (none of them was more special than the very first one. She used to read it over and over):

 Extended Version

# How Deep Does It Go

*Hey honey, I miss you already, and I just left a week ago.*

*I want a better life for both of us, so I'm going to stick in there*

*No matter how much I miss you, I want to come home.*

*Since I've been here, there have been some extremely long days*

*And the only bad part about some of our duties, we still make*

*the same amount no matter how long the day is.*

*I miss you so much; I know I've said that already*

*But, I really do.*

*I can't wait to be back home in your arms*

*I need you to be strong while I'm away,*

*Because I know this is just as hard for you as it is for me.*

*My new commander is a pain in the butt,*

*But I just think about you, and that makes everything better*

*Over here in Iraq, the people live by totally different*

*From what we are used to,*

*There is water that you can't drink and food that you*

*Would never eat,*

*Just seeing how people live here makes me so grateful to live*

*In the United States.*

Extended Version

# Replenishment



*It's definitely an eye-opener and helps you appreciate what you don't have, but I have to go, we are about to surprise Mike*
*for his birthday*
*I love you and miss you.*
*—Jimmy*

Jessica had that letter right in the front of her box, and it gave her hope when she felt like there was no hope. He always had that sort of effect on her since they first met, when he took his coat off and sheltered her from the rain while he got soaking wet just so she could make it to the car and not get her hair wet. He had no idea that she had just gotten her hair done, so that made it even more special to her. He stood there in the rain until she cracked her window and handed him her number. From that moment on, they were inseparable.

******

 Extended Version

# How Deep Does It Go

The doctor had come back in the room with Jessica's test results. "Well, we figured out what's wrong with you," he said.

"What, doctor?" Jessica asked.

The doctor smiled.

"What's going on, doctor?" Jessica asked.

The doctor looked over at Jessica's mother.

"What?" Jessica asked. "Do I have another stomach virus?"

"Jessica, you might want to sit down first."

She took a seat.

"You are pregnant."

Jessica's eyes filled with tears, not for the mere fact that she was pregnant but because Jimmy wasn't there to share the special moment with her. They had been trying to have a child since they got married 2 years ago. She couldn't hold it in. She started crying out, for it was just too emotional at this point in time.

Her mother wrapped her arms around her. "It's OK. Take it easy, honey," she said.

Jessica wept even more.

"Shhhhhh."

Jessica clinched her mother's shirt tightly. The doctor got up and placed his hand on Jessica's shoulder. "I'll leave you two alone for a moment," he said. He walked out and closed the door softly behind him.

Jessica wiped her eyes. "Mom, this is so messed up. We have been trying to have a baby forever, and now that Jimmy's gone, I end up pregnant. This is so hurtful; I don't know what to do."

"Jessica, I know it's hard, but I will be here with you through it all. Hopefully, Officer Dean finds something to find his whereabouts. We can stop by

 Extended Version

my house before we head back to your house. I need to grab some stuff. I'm going to stay with you for a while and help any way I can."

Jessica sniffled. "OK."

Deborah wiped the tears from Jessica's face. Jessica smirked.

"You OK?"

"Yes, I'm fine, thank you, mom. I don't know what I would do without you."

"Good thing you don't have to find out."

Jessica smiled.

The doctor knocked on the door and walked back in the room. "Well, let's get you out of here. I'm going to set you up with some prenatal vitamins. Soon, the sickness will ease up. So, stay active and drink plenty of water so you can stay hydrated.

There are a few side effects that you may encounter when you're pregnant. You may experience tender or swollen breast, nausea with or without vomiting, which you have already experienced–but during the first trimester, increased urination, fatigue, vaginal discharge, and an increase in appetite. If you experience anything other than that, contact the office with any concerns that you may have. Other than that, you are perfectly fine and congratulations."

"OK, thank you, doctor."

"You're welcome, Jessica, and take care."

"OK."

"You can leave the room when you're ready."

"OK, thank you."

The doctor left the room and closed the door behind him. Jessica and her mother got up and left the doctor's office. They both got in the car and headed to Deborah's house. Twenty minutes later, they pulled up and got out and headed into the

 Extended Version

house. Deborah hung up her keys on the key holder by the door and walked upstairs to get some clothes.

"Honey, I just went shopping, so if you want something to drink or something, go ahead and get whatever your heart desires."

"OK, mom. Hopefully, I can hold something down."

"Then again, don't worry about it."

Jessica smirked.

"I'll be down in a minute."

"OK, mom. Take your time."

Deborah walked into her room, grabbed her duffle bag out of the closet and started to pack up a few things.

"You good, honey?"

Jessica had grabbed a picture (off the table) that her mother had of Jimmy and her when they first got married. All she could do was look at it and wish Jimmy was there.

Deborah stopped packing.

"Jessica, are you OK, honey?"

Jessica looked up. "Yes, I'm fine, mother."

"OK, honey."

Jessica took a seat and started to look at the picture again.

"Jimmy, come back to me," she whispered. Suddenly, her phone rang, and she answered. "Hello."

"You have collected call for Phillip. Do you accept the charges?" said the automated system.

"Yes, I do," said Jessica.

There was a slight pause.

"Phillip, are you there?"

"Yes, I'm here."

"So, what's going on? Anything new? How long are they planning on keeping you locked up?"

 Extended Version

"Not sure yet. My lawyer is working on my case. The way things are looking, my lawyer wants me to plead insanity, and then spend time in rehab so they can't lock me up. He's my lawyer, and he doesn't even believe me."

"Wow, if you did plead insanity, how long would you have to do the rehab program?"

"Not sure, but most likely, they would put me on some type of medicine, and I would have to do counseling. I don't think I can do this. I might have to do it. They don't have any hard evidence on me that will stick yet.

Wayne is just considered as missing for now; they don't have enough evidence to blame it on me, and the guys who got killed in the woods were killed by a weapon that hasn't been found and never will because it wasn't either of our weapons that killed them. So, the way things are looking, I have a 50/50 chance right now."

"Well, I will pray for you, and if you need me to do anything, just let me know. I will do everything that I can do for you."

"OK, I will keep you posted. Well, I got to go. Talk to you as soon as I figure out what I want to do."

"OK. Before you leave, I have something to tell you."

"What is it?"

"I went to the doctor today, and you are a grandfather now."

Phillip paused.

"Phillip, are you there?"

"Yes, I'm here. I'm just shocked that I'm going to be a grandfather."

"Yes, you are."

Phillip smirked.

 Extended Version

"I can start calling you grandpops now."

"Hey, don't push it."

Jessica laughed.

"Well, I guess I will just have to stick it out, and hopefully, all this will be over before he's born. Talk to you soon. I'm so excited. Bye-bye …"

Jessica smiled.

Phillip hung up the phone.

******

Deborah had finished packing her bag and had made it back downstairs.

"You're ready, Jessica."

"Yes, I am, mom. Thanks again for coming over to be with me."

Jessica got up and hugged her mother and smiled. Even though Jimmy wasn't there, she still had a loving family until she could fill the void of Jimmy's disappearance.

They left and headed back to Jessica's house.

**** 1 MONTH AND A HALF LATER ****

It was Friday, October 6th. Phillip had been locked up for about 2 months while officer Dean did an intense investigation. During that time, Dean tried his best to gather anything that he could use to prove Phillip guilty. The pickings were very slim; Dean didn't have much to work with besides that Phillip was present on both occasions, which he was about to find out how the shortage of evidence would hold up in court.

Phillip had decided not to plead insanity after he found out that he was going to be a grandfather, so he decided to take his chances in court. His lawyer strayed away from the alien tale and stuck with facts that should prove his innocence.

The case of Phillip and Jonathan had been going on for a few days, and the courtroom had grown completely silent after a long tiring case. It all boiled down to the decision of the jury at this point, so all that was left to do, while the wait progressed, was to pray. Phillip's family had all come out to support him through this process. Surprisingly, his ex-wife even made her way out to the courthouse. She didn't agree with their past issues, but she knew he wasn't a murderer.

Hopefully, fate was on his side; the judge was a good friend of his. They attended hunting trips together, had family gatherings; he even joined them for Christmas occasionally. They had been in the courtroom plenty of times; justifying cases and making sure criminals got what they deserved while innocent citizens sought justice.

Through this whole endeavor, once Phillip lost his son, he had lost all will to get out of jail. Now that he had another reason to be on the outside, he'd been focused on gaining his freedom. Besides, now he had a nephew to protect against the Whiffer. Phillip knew that they would be back, and he would be more prepared. *They wouldn't come back here and take anyone I love,* he thought. *This was their first and last time.*

The jury had finally made their decision, for the judge had come back to the podium, and everyone grew even more quiet as they waited to see what the jury decided. Dean looked over at Phillip and nodded his head, wishing him the best of luck. Everyone stood to his or her feet as the judge entered. When he took a seat, the deputy signaled everyone to take a seat.

          Extended Version

# How Deep Does It Go

"You may all be seated; court is now in session."

The Judge looked around when everyone took a seat; he shuffled through the paper on his bench. "In case number 562, Phillip vs. State, the jury has made their decision. The jury has ruled Phillip, for the murder of Wayne Banks, not guilty.

The courtroom applauded.

"The jury has ruled case 563. Phillip vs. State for the murder of Kevin Burger, Harry Reed, and Thomas Pierce not guilty."

The courtroom stood and applauded loudly.

"Order in the courtroom!" said Judge Jerry. "Order in the courtroom!"

"In case number 585, Jonathan vs. State, for the murder of Kevin Burger, Harry Reed, and Thomas Pierce, not guilty."

The courtroom shouted for joy.

Extended Version 437

# Replenishment

"Court is adjourned."

Phillip jumped and hugged his lawyer and hugged Jonathan.

"We did it, Jonathan."

"Yeah, we made it through; we are finally free."

"Yeah, we can finally go home to our families."

Jonathan looked at Phillip.

"Yeah"

Jonathan gave him a hug again.

Phillip stood there for a moment, thinking about Jimmy.

"Keep your head up."

Phillip was trying to hold himself together. With all the celebration going on around him, all he could do was think about Jimmy at that moment. Dean walked over to him and shook his hand.

"Sorry, we couldn't find your son."

"It's OK; thanks for trying. I appreciate it."

 Extended Version

"You're welcome. See you back at work on Monday."

Phillip nodded.

Phillip walked up to Alice, and she gave him a hug.

"I'm happy you got out."

"Thank you."

"Please, don't ever stop looking for our son."

"I won't, I promise."

Alice hugged him and walked out the courtroom, wiping the tear from her eye.

Phillip walked out of the courthouse with Jessica and Deborah; they went back to Jessica's house.


**** ****

Extended Version

# Chapter 29

The Return

Jimmy had been traveling through space for about a day. He had finally come close enough that he could see the earth. For a second, he got extremely happy, but when he actually thought about it, he was still in alien form and would be hiding for the rest of his life unless he could figure out a way to regain his human form, which was probably impossible.

Jimmy missed his wife and dad so much that he just had to see them, regardless if he could say hello or not. He just wanted to make sure they were OK. He knew his wife probably was taking his disappearance harder than anyone. He knew by now



Extended Version

his wife had told his father what happened; he knew his father wouldn't ever stop looking for him.

He grew close enough to the earth that he could see the water surrounding the land. He stopped the ship and put it in hover mode just so he could look at the planet like he had never seen it before. This was a once in a lifetime experience. Just looking down, he wondered how the planet and human came into existence. It was something that no one could explain. So many different things ran through his head as he hovered out in space just above the earth.

The ship automatically started scanning the planet and displayed different coordinates and locations to land. Jimmy picked somewhere just outside of his house, far enough that no one could see him land. He then took one more glimpse of the earth and the stars that surrounded the earth. He took a deep breath, placed his hand on the controls and proceeded to enter the atmosphere.

The ship pierced the atmosphere and was traveling extremely fast, headed right towards the earth as if it was going to plunge into the earth. He could see his house, as he grew closer to the earth. Suddenly, the ship triggered the landing button, prompting Jimmy to push it. He pushed the button, but the ship didn't slow down at all. It locked in on a location in the field Jimmy had chosen for his location and sped up. Jimmy prepared himself for a crash landing.

The ship was about 200 feet from impact, and Jimmy covered his face and started screaming, for he thought these were his last moments of life. The ship reached 20 feet from the ground and came to a hauling stop and started to hover down to the ground slowly; it activated its cloaking device, which cascaded from the bottom of the ship to the top as it gently sat down on the grass field.

Jimmy slowly opened his eyes. "Thank you, Lord!" He thought he was a goner for sure. The ship's door opened up, and Jimmy stepped out of the ship; he was extremely happy to be home again, but he was disappointed at the same time. He knew he would never be able to talk to his family again. All he could do was go by and see everyone. The first person he went to see was his father. When he left, they were searching for him, so he wanted to see if he had made it back home.

When Jimmy got to his father's house, it looked totally different. He had all types of contraptions up, like he was waiting for something. Jimmy looked through the window and saw him sitting on the couch, and suddenly, Phillip heard an alarm go off, so he looked at the window, and Jimmy ran off; Phillip could hear the wind as he ran off.

Phillip grabbed his shotgun and ran over to the window and looked at the trees to see if they were moving. His alarm shut off automatically once Jimmy was out of reach. Phillip had set up heat sensors, motion detectors, and infrared cameras all around his house.

Once Jimmy got at a far enough distance, he looked back at his father and watched him walk away from the window. "Glad you're OK, Dad." Jimmy then went to his home and saw Jessica sitting in a chair on the porch. His clocking device got activated (it automatically got activated when they were 50 feet away from humans). He walked up to her and went to sniff her and looked over her shoulder and noticed a little boy sitting on a blanket on the porch, playing with toys. He was curious whose child he was.

# How Deep Does It Go

"Jericho are you having fun?" said Jessica.

Jimmy was confused, for that was supposed to be the name of their first baby boy if they could ever have one. Jericho looked up at his mother, and then he looked directly at Jimmy, and one of his eyes transformed into an eye, which looked just like Jimmy's eyes (Whiffer). His son had acquired an alien ability. When Jessica was scratched during Jimmy's seduction, the Whiffer's blood got in her cut and then joined her bloodstream, which gave Jericho special abilities. He could not only see Whiffers; he had acquired other gifts as well.

Jericho pointed right at Jimmy.

Jessica turned around to look.

Jimmy ran around the side of the house.

"Jericho, what are you pointing at, silly boy?"

Jimmy stood beside the house, thinking to himself. *Is this my son? How is this possible? I've only been gone a few*

# Replenishment

*days*. Little did he know that traveling between universes sped up time by 2 years. It had been two years since he had been on earth, but to him, it seemed like a few days.

Jimmy didn't realize it had been 2 years since he had left planet Zara. The Queen laid an egg that wasn't destroyed while the chamber got flooded. The egg sat on the hill, so when the chamber flooded, the water cascaded around the hill and missed the egg completely.

The queen's name is Zara, and its memories were passed down from queen to queen, so all new information was consumed and passed down as well. So, the queen knew everything from the prehistoric times until the present. The egg had just hatched, and the saga continued.



Extended Version

The End

# Replenishment



Extended Version